BASICS OF SEMICONDUCTOR

BASICS OF SEMICONDUCTOR

Alimuddin Khan

ANMOL PUBLICATIONS PVT. LTD.
NEW DELHI - 110 002 (INDIA)

ANMOL PUBLICATIONS PVT. LTD.
Regd. Office: 4360/4, Ansari Road, Daryaganj,
New Delhi-110002 (India)
Tel.: 23278000, 23261597, 23286875, 23255577
Fax: 91-11-23280289
Email: anmolpub@gmail.com
Visit us at: www.anmolpublications.com

Branch Office: No. 1015, Ist Main Road, BSK IIIrd Stage
IIIrd Phase, IIIrd Block, Bengaluru-560 085 (India)
Tel.: 080-41723429 • Fax: 080-26723604
Email: anmolpublicationsbangalore@gmail.com

Basics of Semiconductor

First Published, 2008

Reprint, 2012

ISBN 978-81-261-3563-9

PRINTED IN INDIA

Contents

Preface

This book was motivated by my experience in teaching the Physics course in the Electronics Engineering program at the University of Amravati. Introduction to semiconductor is designed for students, engineers, and scientists seeking a comprehensive background in Modern physics in the minimum time with the minimum number of mathematical and physical concepts.

In Fundamentals of Semiconductor Physics and Devices, AUTHOR have provided a very extensive and detailed text on the physics underlying semiconductor devices. More so than any other current text, this book provides a greatly expanded discussion of modern tight-binding methods, helping the students to understand these aspects of electronic structure in clear, simple terms. In connection with this the authors offer a very detailed discussion of deep levels in semiconductors, which are so important to semiconducting properties. Also, in the discussion of transport properties, the book goes into much greater depth about nonlinear and nonequilibrium processes than is usual. It is quite a unique contribution, containing the basic physics which tends to be missing from device-oriented books, but going much further into the essentials needed for device development than any solid-state-physics text.

Author

Chapter 1

Semiconductor Fundamentals

INTRODUCTION

To understand the fundamental concepts of semiconductors, one must apply modern physics to solid materials. More specifically, we are interested in semiconductor crystals. Crystals are solid materials consisting of atoms, which are placed in a highly ordered structure called a lattice. Such a structure yields a periodic potential throughout the material, which results in some remarkable properties.

Two properties of crystals are of particular interest, since they are needed to calculate the current in a semiconductor. First, we need to know how many fixed and mobile charges are present in the material. Second, we need to understand the transport of the mobile carriers through the semiconductor.

In this chapter we start from the atomic structure of semiconductors and explain the concepts of energy bands, energy band gaps and the density of states in an energy band. We also show how the current in an almost filled band can more easily be analyzed using the concept of holes. Next, we discuss the probability that energy levels within an energy band are occupied. We will use this probability density to find the density of electrons and holes in a band.

Two carrier transport mechanisms will be considered. The drift of carriers in an electric field and the diffusion of carriers due to a carrier density gradient will be discussed. Recombination mechanisms and the continuity equations are then combined into the diffusion equation. Finally, we present the drift-diffusion model, which combines all the essential elements discussed in this chapter.

CRYSTALS AND CRYSTAL STRUCTURES

Solid materials are classified by the way the atoms are arranged within the solid. Materials in which atoms are placed at random are called amorphous. Materials in which atoms are placed in a high ordered structure are called crystalline. Poly-crystalline materials are materials with a high degree of short-range order and no long-range order. These materials consist of small crystalline regions with random orientation called grains, separated by grain boundaries.

Of primary interest in this text are crystalline semiconductors in which atoms are placed in a highly ordered structure. Crystals are categorized by their crystal structure and the underlying lattice. While some crystals have a single atom placed at each lattice point, most crystals have a combination of atoms associated with each lattice point. This combination of atoms is also called the basis.

The classification of lattices, the common semiconductor crystal structures and the growth of single-crystal semiconductors are discussed in the following sections.

BRAVAIS LATTICES

The Bravais lattices are the distinct lattice types, which when repeated can fill the whole space. The lattice can therefore be generated by three unit vectors, $\vec{a}_1, \vec{a}_2$ and $\vec{a}_3$ and a set of integers k, l and m so that each lattice point, identified by a vector $\vec{r}$, can be obtained from:

$$\vec{r} = k\vec{a}_1 + l\vec{a}_2 + m\vec{a}_3$$

The construction of the lattice points based on a set of unit vectors is illustrated by Figure.

In two dimensions, there are five distinct Bravais lattices, while in three dimensions there are fourteen.

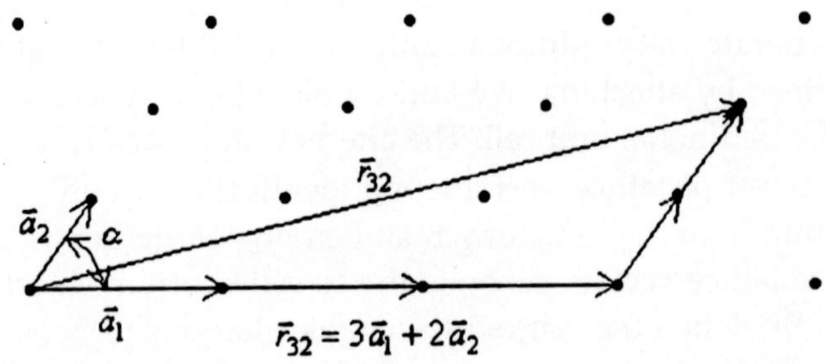

Fig. The construction of lattice points using unit vectors

The lattices in two dimensions are the square lattice, the rectangular lattice, the centred rectangular lattice, the hexagonal lattice and the oblique lattice as shown in Figure.It is customary to organize these lattices in groups, which have the same symmetry. An example is the rectangular and the centred rectangular lattice. As can be seen on the figure, all the lattice points of the rectangular lattice can be obtained by a combination of the lattice vectors $\bar{a}_1$ and $\bar{a}_2$.

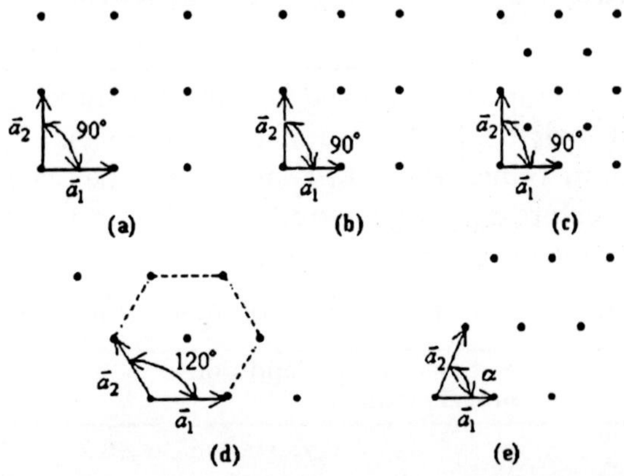

Fig. The five Bravais lattices of two-dimensional crystals: (a) square, (b) rectangular, (c) centred rectangular, (d) hexagonal and (e) oblique

The centred rectangular lattice can be constructed in two ways. It can be obtained by starting with the same lattice vectors as those of the rectangular lattice and then adding an additional atom at the centre of each rectangle in the lattice. This approach is illustrated by Figure. The lattice vectors

$\bar{a}_1$ generate the traditional unit cell and the centre atom is obtained by attaching two lattice points to every lattice point of the traditional unit cell. The alternate approach is to define a new set of lattice vectors, one identical to $\bar{a}_1$ and another starting from the same origin and ending on the centre atom. These lattice vectors generate the so-called primitive cell and directly define the centred rectangular lattice.

These lattices are listed in Table. a_1 and a_2 are the magnitudes of the unit vectors and α is the angle between them.

Table: Bravais lattices of two-dimensional crystals

Name	Number of Bravais lattices	Conditions
Square	1	$\alpha_1 = \alpha_2, \alpha = 90°$
Rectangular	2	$\alpha_1 \neq \alpha_2, \alpha = 90°$
Hexagonal	1	$\alpha_1 = \alpha_2, \alpha = 120°$
Oblique	1	$\alpha_1 \neq \alpha_2, \alpha \neq 120° \alpha \neq 90°$

The same approach is used for lattices in three dimensions. The fourteen lattices of three-dimensional crystals are classified as shown in Table, where a_1, a_2 and a_3 are the magnitudes of the unit vectors defining the traditional unit cell and α, β and γ are the angles between these unit vectors.

Table .: Bravais lattices of three-dimensional crystals

Name	Number of Bravais lattices	Conditions
Triclinic	1	$\alpha_1 \neq \alpha_2 \neq \alpha_3, \alpha \neq \beta \neq \gamma$
Monolinic	2	$\alpha_1 \neq \alpha_2 \neq \alpha_3, \alpha = \beta = 90° \neq \gamma$
Orthorhombic	4	$\alpha_1 \neq \alpha_2 \neq \alpha_3, \alpha = \beta = \gamma = 90°$
Tetragonal	2	$\alpha_1 = \alpha_2 \neq \alpha_3, \alpha = \beta = \gamma = 90°$
Cubic	3	$\alpha_1 = \alpha_2 = \alpha_3, \alpha = \beta = \gamma = 90°$
Trigonal	1	$\alpha_1 = \alpha_2 = \alpha_3, \alpha = \beta = \gamma < 120° \neq 90°$
Hexagonal	1	$\alpha_1 = \alpha_2 \neq \alpha_3, \alpha = \beta = 90°, \gamma = 120°$

The cubic lattices are an important subset of these fourteen Bravais lattices since a large number of semiconductors are cubic. The three cubic Bravais lattices are the simple cubic lattice, the body-centred cubic lattice and the face-centred cubic lattice as shown in Figure. Since all unit vectors identifying the traditional unit cell have the same size, the crystal structure is completely defined by a single number. This number is the lattice constant, *a*.

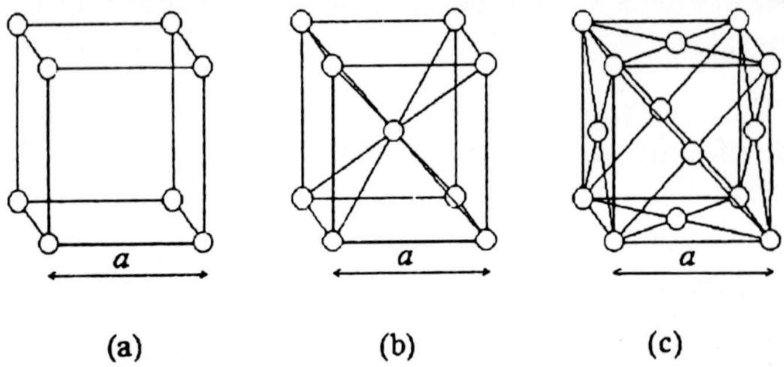

Fig. The simple cubic (a), the body-centred cubic (b) and the face centred cubic (c) lattice.

MILLER INDICES, CRYSTAL PLANES AND DIRECTIONS

Crystal planes of a crystal are characterized by their Miller indices. The Miller indices are defined as the smallest possible integers, which have the same ratios as the inverse of the intersections of a given plane with a set of axis defined by the unit vectors of that crystal. This definition is further illustrated with Figure.The intersections between the plane and the axis occur at p, q, and r.

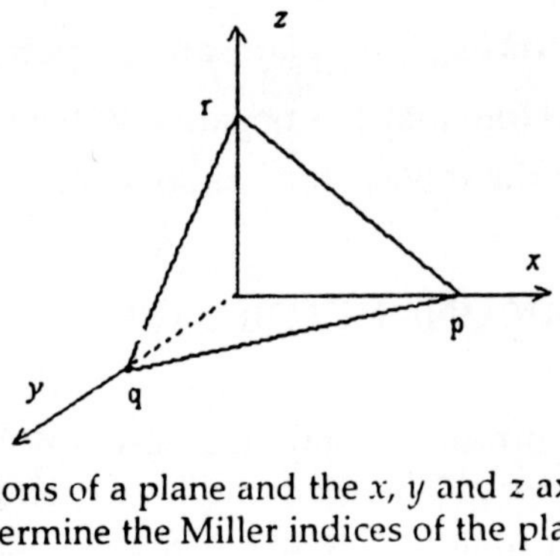

Fig. Intersections of a plane and the *x*, *y* and *z* axis, as used to determine the Miller indices of the plane.

The corresponding Miller indices are therefore $\left(\frac{A}{p}\frac{A}{q}\frac{A}{r}\right)$, where A is an integer chosen such that the Miller indices are the smallest possible integers. It should be clear to the reader that the resulting Miller indices are the same for all parallel planes of atoms in a crystal.

Example

The intercepts of a crystal plane with the axis defined by a set of unit vectors are at 2a, -3b and 4c. Find the corresponding Miller indices of this and all other crystal planes parallel to this plane.

Solution

The Miller indices are obtained in the following three steps:

1. Identify the intersections with the axis, namely 2, -3 and 4
2. Calculate the inverse of each of those intercepts, resulting in 1/2, –1/3 and 1/4.
3. Find the smallest integers proportional to the inverse of the intercepts. Multiplying each fraction with the product of each of the intercepts (24 = 2 × 3 × 4) does result in integers, but not always the smallest integers. These are obtained in this case by multiplying each fraction by 12.

The Miller indices of this plane and all parallel planes are therefore $(6\bar{4}3)$. Note that the negative indices are indicated with a bar above the integer for a more compact notation.

COMMON SEMICONDUCTOR CRYSTAL STRUCTURES

The most common crystal structure among frequently used semiconductors is the diamond lattice, shown in Figure. Each atom in the diamond lattice has a covalent bond with four adjacent atoms, which together form a tetrahedron. This lattice can also be formed from two face-centred-cubic lattices, which are displaced along the body diagonal of the larger cube in Figure by one quarter of that body diagonal. The diamond lattice therefore is a face-centred-cubic lattice with a basis containing two identical atoms.

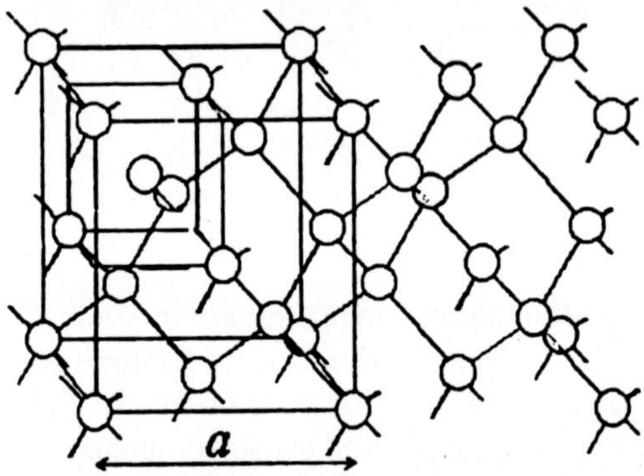

Fig. The diamond lattice of silicon and germanium

Compound semiconductors such as GaAs and InP have a crystal structure that is similar to that of diamond. However, the lattice contains two different types of atoms. Each atom still has four covalent bonds, but they are bonds with atoms of the other type. This structure is referred to as the zinc-blende lattice, named after zinc-blende (ZnS) as shown in Figure. Both the diamond lattice and the zinc-blende lattice are cubic lattices. A third common crystal structure is the hexagonal structure also referred to as the wurzite crystal structure, which is the hexagonal form of zinc sulfide (ZnS).

Many semiconductor materials can have more than one crystal structure. A large number of compound semiconductors including GaAs, GaN and ZnS can be either cubic or hexagonal. SiC can be cubic or one of several different hexagonal crystal structures.

The cubic crystals are characterized by a single parameter, the lattice constant *a*, while the hexagonal structures are characterized in the hexagonal plane by a lattice constant *a* and by the distance between the hexagonal planes, *c*.

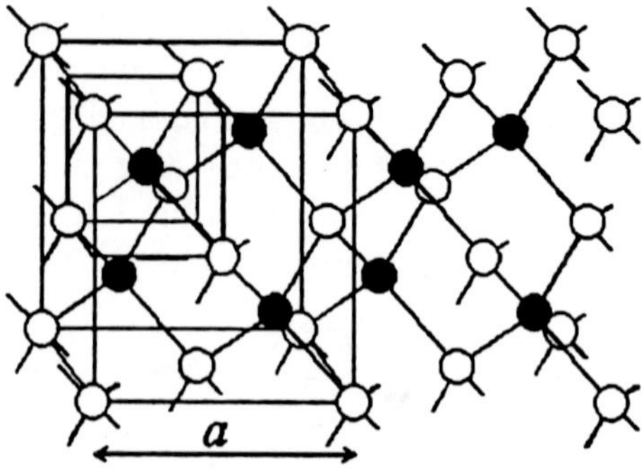

Fig. The zinc-blende crystal structure of GaAs and InP

Example

Calculate the maximum fraction of the volume in a simple cubic crystal occupied by the atoms. Assume that the atoms are closely packed and that they can be treated as hard spheres. This fraction is also called the packing density.

Solution

The atoms in a simple cubic crystal are located at the corners of the units cell, a cube with side *a*. Adjacent atoms touch each other so that the radius of each atom equals *a*/2. There are eight atoms occupying the corners of the cube, but only one eighth of each is within the unit cell so that the number of atoms equals one per unit cell. The packing density is then obtained from:

$$\frac{\text{Volume of atoms}}{\text{Volume of the unit cell}} = \frac{\frac{4}{3}\pi r^3}{a^3} = \frac{\frac{4}{3}\pi(\frac{a}{2})^3}{a^3} = \frac{\pi}{6} = 52\%$$

or about half the volume of the unit cell is occupied by the atoms.

The packing density of four cubic crystals is listed in the table below.

	Radius	Atoms/ unit cell	Packing density
Simple cubic	$\frac{a}{2}$	1	$\frac{\pi}{6} = 52\%$

Body centered cubic	$\frac{\sqrt{3}a}{4}$	2	$\frac{\pi\sqrt{3}}{8} = 68\%$
Face centered cubic	$\frac{\sqrt{2}a}{4}$	4	$\frac{\pi\sqrt{3}}{16} = 34\%$
Diamond	$\frac{\sqrt{3}a}{8}$		$\frac{\pi\sqrt{3}}{16} = 34\%$

GROWTH OF SEMICONDUCTOR CRYSTALS

Like all crystals, semiconductor crystals can be obtained by cooling the molten semiconductor material. However, this procedure yields poly-crystalline material since crystals start growing in different locations with a different orientation. Instead when growing single-crystalline silicon one starts with a seed crystal and dips one end into the melt. By controlling the temperature difference between the seed crystal and the molten silicon, the seed crystal slowly grows. The result is a large single-crystal silicon boule. Such boules have a cylindrical shape, in part because the seed crystal is rotated during growth and in part because of the cylindrical shape of the crucible containing the melt. The boule is then cut into wafers with a diamond saw and further polished to yield the starting material for silicon device fabrication.

ENERGY BANDS

Energy bands consisting of a large number of closely spaced energy levels exist in crystalline materials. The bands can be thought of as the collection of the individual energy levels of electrons surrounding each atom. The wavefunctions of the individual electrons, however, overlap with those of electrons confined to neighbouring atoms. The Pauli exclusion principle does not allow the electron energy levels to be the same so that one obtains a set of closely spaced energy levels, forming an energy band. The energy band model is crucial to any detailed treatment of semiconductor devices. It provides the framework needed to understand the concept of an energy bandgap and that of conduction in an almost filled band as described by the empty states.

FREE ELECTRON MODEL

The free electron model of metals has been used to explain the photo-electric effect. This model assumes that electrons are free to move within the metal but are confined to the metal by potential barriers as illustrated by Figure. The minimum energy needed to extract an electron from the metal equals $q\Phi_M$, where Φ_M is the workfunction. This model is frequently used when analyzing metals. However, this model does not work well for semiconductors since the effect of the periodic potential due to the atoms in the crystal has been ignored.

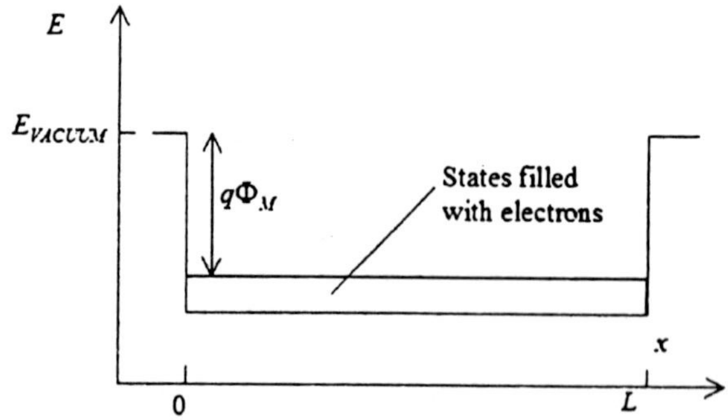

Fig. The free electron model of a metal.

PERIODIC POTENTIALS

The analysis of periodic potentials is required to find the energy levels in a semiconductor. This requires the use of periodic wave functions, called Bloch functions which are beyond the scope of this text. The result of this analysis is that the energy levels are grouped in bands,The analysis of periodic potentials is required to find the energy levels in a semiconductor. This requires the use of periodic wave functions, called Bloch functions which are beyond the scope of this text. The result of this analysis is that the energy levels are grouped in bands, separated by energy band gaps. The behaviour of electrons at the top and bottom of such a band is similar to that of a free electron. However, the electrons are affected by the presence of the periodic potential. The combined effect of the periodic potential is included by adjusting the mass of the electron to a different value. This mass will be referred to as the effective mass. separated by energy band gaps. The behaviour of electrons at the top and bottom of such a band is similar to that of a free electron. However, the electrons are affected by the presence of the periodic potential. The combined effect of the periodic potential is included by adjusting the mass of the electron to a different value. This mass will be referred to as the effective mass.

The effect of a periodic arrangement on the electron energy levels is illustrated by Figure. Shown are the energy levels of electrons in a carbon crystal with the atoms arranged in a diamond lattice. These energy levels are plotted as a function of the lattice constant, *a*.

Energy bands for diamond versus lattice constant. One atomic unit equals 1 Rydberg = 13.6 eV.

Isolated carbon atoms contain six electrons, which occupy the 1s, 2s and 2p orbital in pairs. The energy of an electron occupying the 2s and 2p orbital is indicated on the figure. The energy of the 1s orbital is not shown. As the lattice constant is reduced, there is an overlap of the electron wavefunctions occupying adjacent atoms. This leads to a splitting of the energy levels consistent with the Pauli exclusion principle. The splitting results in an energy band containing 2N states in the 2s band and 6N states in the 2p band, where N is the number of atoms in the crystal. A further reduction of the lattice constant causes the 2s and 2p energy bands to merge and split again into two bands containing 4N states each. At zero Kelvin, the lower band is completely filled with electrons and labeled as the valence band. The upper band is empty and labeled as the conduction band.

ENERGY BANDS OF SEMICONDUCTORS

Complete energy band diagrams of semiconductors are very complex. However, most have features similar to that of the diamond crystal. We first take a closer look at the energy band diagrams of common semiconductors. We then present a simple diagram containing some of the most important feature and discuss the temperature dependence of the energy bandgap.

Energy Band Diagrams of Common Semiconductors

The energy band diagrams of semiconductors are rather complex. The detailed energy band diagrams of germanium, silicon and gallium arsenide are shown in Figure. The energy is plotted as a function of the wavenumber, k, along the main crystallographic directions in the crystal, since the band diagram depends on the direction in the crystal. The energy band diagrams contain multiple completely-filled and completely-empty bands. In addition, there are multiple partially-filled band.

Fortunately, we can simplify the energy band diagram since only the electrons in the highest almost-filled band and the lowest almost-empty band dominate the behaviour of the semiconductor. These bands are indicated on the figure by the + and - signs corresponding to the charge of the carriers in those bands.

Simple Energy Band Diagram of a Semiconductor

The energy band diagrams shown in the previous section are frequently simplified when analyzing semiconductor devices. Since the electronic properties of a semiconductor are dominated by the highest partially empty band and the lowest partially filled band, it is often sufficient to only consider those bands. This leads to a simplified energy band diagram for semiconductors as shown in Figure:

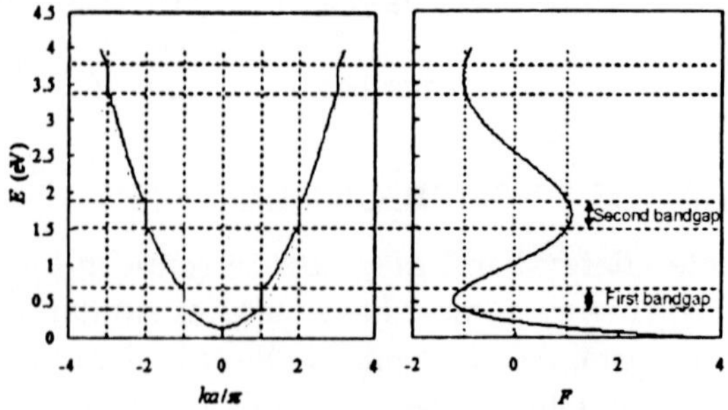

Fig. Energy Band

A simplified energy band diagram used to describe semiconductors. Shown are the valence and conduction band as indicated by the valence band edge, E_v, and the conduction band edge, E_c. The vacuum level, E_{vacuum}, and the electron affinity, χ, are also indicated on the figure.

The diagram identifies the almost-empty *conduction band* by a horizontal line. This line indicates the bottom edge of the conduction band and is labeled E_c. Similarly, the top of the valence band is indicated by a horizontal line labeled E_v. The energy band gap is located between the two lines, which are separated by the bandgap energy E_g. The distance between the conduction band edge, E_c, and the energy of a free electron outside the crystal (called the vacuum level labeled E_{vacuum}) is quantified by the electron affinity, χ multiplied with the electronic charge q.

An important feature of an energy band diagram, which is not included on the simplified diagram, is whether the conduction band minimum and the valence band maximum occur at the same value for the wavenumber. If so, the energy bandgap is called direct. If not, the energy bandgap is called indirect. This distinction is of interest for optoelectronic devices as direct bandgap materials provide more efficient absorption and emission of light. For instance, the smallest bandgap of germanium and silicon is indirect, while gallium arsenide has a direct bandgap as can be seen on Figure.

Temperature Dependence of the Energy Bandgap

The energy bandgap of semiconductors tends to decrease as the temperature is increased. This behaviour can be better understood if one considers that the interatomic spacing increases when the amplitude of the atomic vibrations increases due to the increased thermal energy. This effect is quantified by the linear expansion coefficient of a material. An increased interatomic spacing decreases the average potential seen by the electrons in the material, which in turn reduces the size of the energy bandgap. A direct modulation of the interatomic distance - such as by applying compressive (tensile) stress - also causes an increase (decrease) of the bandgap.

The temperature dependence of the energy bandgap, E_g, has been experimentally determined yielding the following expression for E_g as a function of the temperature, *T*:

$$\cos ka = F = \frac{\alpha^2 - \beta^2}{2\alpha\beta} \sinh \alpha b \sin \beta(a-b) + \cosh \alpha b \cos \beta(a-b)$$

where $E_g(0)$, α and β are the fitting parameters. These fitting parameters are listed for germanium, silicon and gallium arsenide in Table:

Table

	Germanium	Silicon	GaAs
$E_g(0)$(eV)	0.7437	1.166	1.519
α(meV/K)	0.477	0.473	0.541
β(K)	235	636	204

Parameters used to calculate the energy bandgap of germanium, silicon and gallium arsenide (GaAs) as a function of temperature

A plot of the resulting bandgap versus temperature is shown in Figure for germanium, silicon and gallium arsenide.

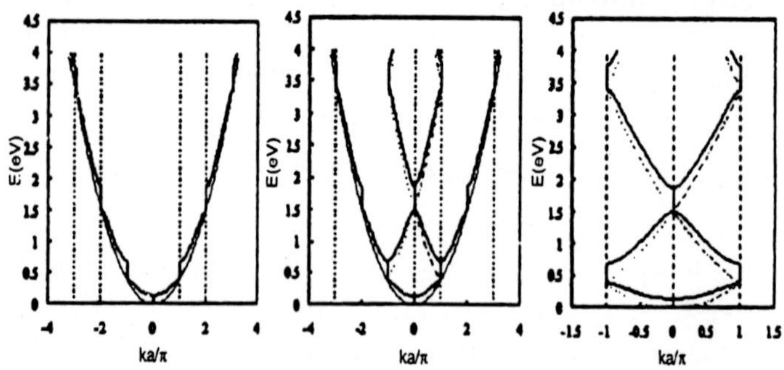

Fig. Temperature dependence of the energy bandgap of germanium (Ge), silicon (Si) and gallium arsenide (GaAs).

Example .

Calculate the energy bandgap of germanium, silicon and gallium arsenide at 300, 400, 500 and 600 K.

Solution

The bandgap of silicon at 300 K equals:

$$\text{Eg(300k)}= E_g(0K) - \frac{\alpha T^2}{T+\beta} = 1.166 - \frac{0.473\times(300)^2}{300+636} = 1.12eV$$

Similarly one finds the energy bandgap for germanium and gallium arsenide, as well as at different temperatures, yielding:

	Germanium	Silicon	Gallium Arsenide
T = 300K	0.66eV	1.12eV	1.42eV
T = 400K	0.62eV	1.09eV	1.38eV
T = 500K	0.58eV	1.06eV	1.33eV
T = 600K	0.54eV	1.03eV	1.28eV

Doping Dependence of the Energy Bandgap

High doping densities cause the bandgap to shrink. This effect is explained by the fact that the wavefunctions of the electrons bound to the impurity atoms start to overlap as the density of the impurities increase. For instance, at a doping density of 1018 cm-3, the average distance between two impurities is only 10 nm. This overlap forces the energies to form an energy band rather than a discreet level.

If the impurity level is shallow, this impurity band reduces the energy band of the host material by:

$$\alpha = \frac{\sqrt{2m(V_0 - E)}}{\hbar} \text{ and } \beta = \frac{\sqrt{2mE}}{\hbar}$$

where N is the doping density, q is the electronic charge, es is the dielectric constant of the semiconductor, k is Boltzmann's constant and T is the temperature in Kelvin. For silicon (er = 11.7) this expression further reduces to:

$$\cos ka = F = p\frac{\sin \beta a}{\beta a} + \cos \beta a$$

From this expression we find that the bandgap shrinkage can typically be ignored for doping densities less than 1018 cm^{-3}.

A plot of the change in bandgap energy with doping density is shown in Figure:

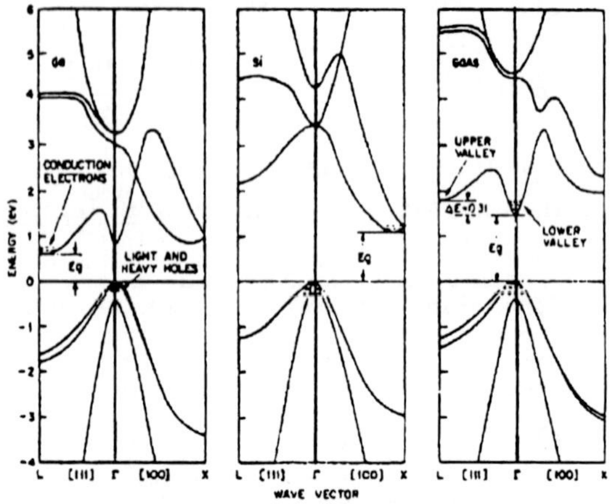

Fig. Doping dependence of the energy bandgap of GaAs (top/red curve), germanium (black curve) and silicon (bottom/blue curve).

METALS, INSULATORS AND SEMICONDUCTORS

Once we know the bandstructure of a given material we still need to find out which energy levels are occupied and whether specific bands are empty, partially filled or completely filled.

Empty bands do not contain electrons. Therefore, they are not expected to contribute to the electrical conductivity of the material. Partially filled bands do contain electrons as well as available energy levels at slightly higher energies. These unoccupied energy levels enable carriers to gain energy when moving in an applied electric field. Electrons in a partially filled band therefore do contribute to the electrical conductivity of the material.

Completely filled bands do contain plenty of electrons but do not contribute to the conductivity of the material. This is because the electrons cannot gain energy since all energy levels are already filled.

In order to find the filled and empty bands we must find out how many electrons can be placed in each band and how many electrons are available. Each band is formed due to the splitting of one or more atomic energy levels. Therefore, the minimum number of states in a band equals twice the number of atoms in the material. The reason for the factor of two is that every energy level can contain two electrons with opposite spin.

To further simplify the analysis, we assume that only the *valence* electrons (the electrons in the outer shell) are of interest. The core electrons are tightly bound to the atom and are not allowed to freely move in the material.

Possible energy band diagrams of a crystal. Shown are

a. A half filled band,

b. Two overlapping bands,

c. An almost full band separated by a small bandgap from an almost empty band and

d. A full band and an empty band separated by a large bandgap.

A half-filled band is shown in Figure. This situation occurs in materials consisting of atoms, which contain only one valence electron per atom. Most highly conducting metals including copper, gold and silver satisfy this condition. Materials consisting of atoms that contain two valence electrons can still be highly conducting if the resulting filled band overlaps with an empty band. This scenario is shown in b). No conduction is expected for scenario d) where a completely filled band is separated from the next higher empty band by a larger energy gap. Such materials behave as insulators. Finally, scenario c) depicts the situation in a semiconductor. The completely filled band is now close enough to the next higher empty band that electrons can make it into the next higher band. This yields an almost full band below an almost empty band. We will call the almost full band the *valence band* since it is occupied by valence electrons. The almost empty band will be called the *conduction band*, as electrons are free to move in this band and contribute to the conduction of the material.

ELECTRONS AND HOLES IN SEMICONDUCTORS

Semiconductors differ from metals and insulators by the fact that they contain an "almost-empty" conduction band and an "almost-full" valence band. This also means that we will have to deal with the transport of carriers in both bands.

To facilitate the discussion of the transport in the "almost-full" valence band of a semiconductor, we will introduce the concept of holes. It is important for the reader to understand that one could deal with only electrons if one is willing to keep track of all the electrons in the "almost-full" valence band. After all, electrons are the only real particles available in a semiconductor.

The concepts of holes is introduced in semiconductors since it is easier to keep track of the missing electrons in an "almost-full" band, rather than keeping track of the actual electrons in that band. We will now first explain the concept of a hole and then point out how the hole concept simplifies the analysis.

Holes are missing electrons. They behave as particles with the same properties as the electrons would have when occupying the same states except that they carry a positive charge. This definition is illustrated further with Figure, which presents the energy band diagram in the presence of an electric field.

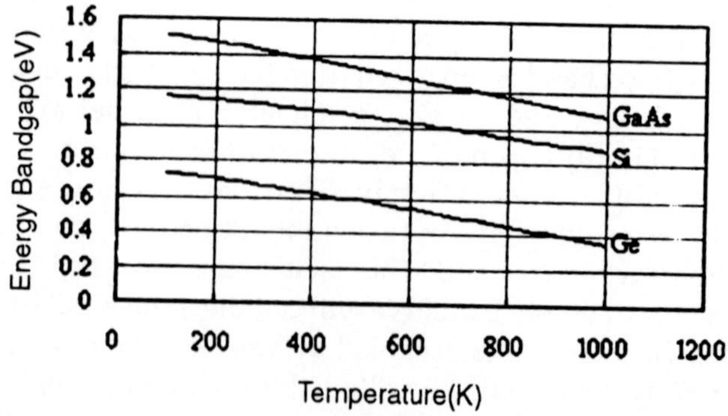

Figure. Energy band diagram in the presence of a uniform electric field. Shown are the upper almost-empty band and the lower almost-filled band. The tilt of the bands is caused by an externally applied electric field.

A uniform electric field is assumed which causes a constant gradient of the bands.

The electrons in the almost-empty band are negatively charged particles, which therefore move in a direction, which opposes the direction of the field. Electrons therefore move *down hill* in the upper band. Electrons in the lower band also move in the same direction. The total current density due to the electrons in the valence band can therefore be written as:

$$\beta = \frac{\sqrt{2mE}}{\hbar} \text{ and } P = \frac{mV_0 ba}{\hbar^2}$$

where V is the volume of the semiconductor, q is the electronic charge and v is the electron velocity. The sum is taken over all occupied or filled states in the lower band. This equation can be reformulated by first taking the sum over all the states in the lower band and subtracting the current due to the electrons, which are missing in the almost-filled band. This last term therefore represents the sum taken over all the empty states in the lower band, or:

$$E_g(T) = E_g(0) - \frac{\alpha T^2}{T + \beta}$$

The sum over all the states in the lower band has to equal zero since electrons in a completely filled band do not contribute to current, while the remaining term can be written as:

$$\Delta E_g(N) = -\frac{3q^2}{16\pi\xi}\sqrt{\frac{q^2 N}{\xi}}$$

which states that the current is due to positively charged particles associated with the empty states in the almost-filled band. We call these particles *holes*. Keep in mind that there is no real particle associated with a hole. Instead, the combined behaviour of all the electrons, which occupy states in the almost-filled band, is the same as that of positively charge particles associated with the unoccupied states.

The reason the concept of holes simplifies the analysis is that the density of states function of a whole band can be rather complex. However, it can be dramatically simplified if only states close to the band edge need to be considered.

THE EFFECTIVE MASS CONCEPT

Electrons with an energy close to a band minimum behave as free electrons. They accelerate in an applied electric field just like a free electron in vacuum. Their wavefunctions are periodic and extend over the size of the material. The presence of the periodic potential, due to the atoms in the crystal without the valence electrons, changes the properties of the electrons. Therefore, the mass of the electron differs from the free electron mass, m_0. Because of the anisotropy of the effective mass and the presence of multiple equivalent band minima, we define two types of effective mass, the effective mass for density of states calculations and the effective mass for conductivity calculations. The effective mass values for electrons and holes are listed together with the value of the smallest energy bandgap in Table. Electrons in gallium arsenide have an isotropic effective mass so that the conductivity effective mass equals the density of states effective mass.

Table

		Germanium	Silicon	GaAs
Smallest energy bandgap at 300K	Eg(eV)	0.66	1.12	1.424
Electron effective mass for density of states calculations	$\frac{m_e^*, dos}{m_0}$	0.55	1.08	0.067
Hole effective mass for density of states calculations	$\frac{m_h^*, dos}{m_0}$	0.37	0.811	0.45

Electron effective mass for conductivity calculations	$\frac{m_e^*, cond}{m_0}$	0.12	0.26	0.067
Hole effective mass for conductivity calculations	$\frac{m_h^*, cond}{m_0}$	0.21	0.386	0.34

Effective mass of carriers in germanium, silicon and gallium arsenide (GaAs)

DETAILED DESCRIPTION OF THE EFFECTIVE MASS CONCEPT

Introduction

The effective mass of a semiconductor is obtained by fitting the actual *E-k* diagram around the conduction band minimum or the valence band maximum by a paraboloid. While this concept is simple enough, the issue turns out to be substantially more complex due to the multitude and the occasional anisotropy of the minima and maxima. In this section we first describe the different relevant band minima and maxima, present the numeric values for germanium, silicon and gallium arsenide and introduce the effective mass for density of states calculations and the effective mass for conductivity calculations.

Most semiconductors can be described as having one band minimum at k = 0 as well as several equivalent anisotropic band minima at $k \neq 0$. In addition there are three band maxima of interest close to the valence band edge.

Band Structure of Silicon

As an example we consider the band structure of silicon as shown in the figure below:

Shown is the *E-k* diagram within the first brillouin zone and along the (100) direction. The energy is chosen to be to zero at the edge of the valence band. The lowest band minimum at k = 0 and still above the valence band edge occurs at $E_{c,direct}$ = 3.2 eV. This is not the lowest minimum above the valence band edge since there are also 6 equivalent minima at k = (x,0,0), (-x,0,0), (0,x,0), (0,-x,0), (0,0,x), and (0,0,-x) with x = 5 nm^{-1}. The minimum energy of all these minima equals 1.12 eV = $E_{c,indirect}$. The effective mass of these anisotropic minima is characterized by a longitudinal mass along the corresponding equivalent (100) direction and two transverse masses in the plane perpendicular to the longitudinal direction. In silicon the longitudinal electron mass is $m_{e,l}^* = 0.98\ m_0$ and the transverse electron masses are $m_{e,t}^* = 0.19\ m_0$, where $m_0 = 9.11 \times 10^{-31}$ kg is the free electron rest mass.

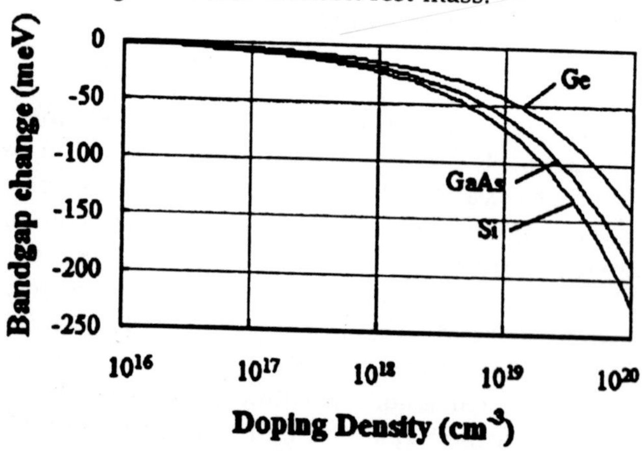

Fig. Simplified E-k diagram of silicon.

Two of the three band maxima occur at 0 eV. These bands are referred to as the light and heavy hole bands with a light hole mass of $m_{l,h}^* = 0.16\ m_0$ and a heavy hole mass of $m_{h,h}^* = 0.46\ m_0$. In addition there is a split-off hole band with its maximum at $E_{v,so} = -0.044$ eV and a split-off hole mass of $m_{h,so}^* = 0.29\ m_0$.

Effective Mass and Energy Band Minima and Maxima of Ge, Si and GaAs

The values of the energy band minima and maxima as well as the effective masses for germanium, silicon and gallium arsenide are listed in the table below:

Name	Symbol	Germanium	Silicon	Gallium Arsenide
Band minimum at k = 0				
Minimum energy	$E_{g,direct}$(eV)	0.8	3.2	1.424
Effective mass	$m_e^*/m0$	0.041	0.2	0.067
Band minimum bot at k = 0				
Minimum energy	$E_{g,\ indirect}$(eV)	0.66	1.12	1.734
Longitudinal effective mass	$m_{e,}I^*/m_0$	1.64	0.98	1.98
Transverse effective mass	$m_{e,}I^*/m_0$	0.082	0.19	0.37
Longitudinal direction		-111	-100	-111
Heavy hole valence band maximum at E = k = 0				
Effective mass	m_{hh}^*/m_0	0.28	0.49	0.45
Light hole valence band maximum at k = 0				
Effective mass	m_{lh}^*/m_0	0.044	0.16	0.082
Split-off hole valence band maximum at k = 0				
Split-off hole valence band energy	$E_{v,so}$(eV)	-0.028	-0.044	-0.34
Effective mass	$m_{h,so}^*/m_0$	0.084	0.29	0.154

Effective mass of carriers in germanium, silicon and gallium arsenide (GaAs)

Effective Mass for Density of States Calculations

The effective mass for density of states calculations equals the mass which provides the density of states using the expression for one isotropic maximum or minimum or:

for the density of states in the conduction band and:

for the density of states in the valence band.

For instance, for a single band minimum described by a longitudinal mass and two transverse masses the effective mass for density of states calculations is the geometric mean of the three masses. Including the fact that there are several equivalent minima at the same energy one obtains the effective mass for density of states calculations from:

$$J_{vb} = \frac{1}{V}\left(\sum_{\substack{\text{all}\\\text{states}}}(-q)v_i - \sum_{\substack{\text{empty}\\\text{states}}}(-q)v_i \right)$$

where M_c is the number of equivalent band minima. For silicon one obtains:

Effective Mass for Conductivity Calculations

The effective mass for conductivity calculation is the mass, which is used in conduction related problems accounting for the detailed structure of the semiconductor. These calculations include mobility and diffusion constants calculations. Another example is the calculation of the shallow impurity levels using a hydrogen-like model.

As the conductivity of a material is inversionally proportional to the effective masses, one finds that the conductivity due to multiple band maxima or minima is proportional to the sum of the inverse of the individual masses, multiplied by the density of carriers in each band, as each maximum or minimum adds to the overall conductivity. For anisotropic minima containing one longitudinal and two transverse effective masses one has to sum over the effective masses in the different minima along the equivalent directions. The resulting effective mass for bands, which have ellipsoidal constant energy surfaces, is given by:

$$g_{,}(E) = \frac{8\pi\sqrt{2}}{h^3} me^{3/2} \sqrt{E - E_e} \text{ 'for } E \geq E_e$$

provided the material has an isotropic conductivity as is the case for cubic materials. For instance electrons in the X minima of silicon have an effective conductivity mass given by:

DENSITY OF STATES

Before we can calculate the density of carriers in a semiconductor, we have to find the number of available states at each energy. The number of electrons at each energy is then obtained by multiplying the number of states with the probability that a state is occupied by an electron. Since the number of energy levels is very large and dependent on the size of the semiconductor, we will calculate the number of states per unit energy and per unit volume.

CALCULATION OF THE DENSITY OF STATES

The density of states in a semiconductor equals the density per unit volume and energy of the number of solutions to Schrödinger's equation. We will assume that the semiconductor can be modeled as an infinite quantum well in which electrons with effective mass, m^*, are free to move. The energy in the well is set to zero. The semiconductor is assumed a cube with side L. This assumption does not affect the result since the density of states per unit volume should not depend on the actual size or shape of the semiconductor.

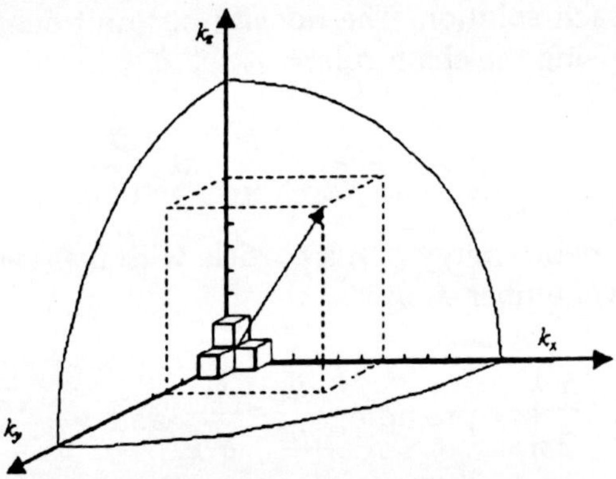

Fig. Calculation of the number of states with wavenumber less than k. The total number of solutions with a different value for k_x, k_y and k_z and with a magnitude of the wavevector less than k is obtained by calculating the volume of one eighth of a sphere with radius k and dividing it by the volume corresponding to a single solution, $\left(\frac{\pi}{L}\right)^3$, yielding:

The solutions to the wave equation where $V(x) = 0$ are sine and cosine functions:

$$\psi = A\sin(k_x x) + B\cos(k_x x)$$

Where A and B are to be determined. The wavefunction must be zero at the infinite barriers of the well. At $x = 0$ the wavefunction must be zero so that only sine functions can be valid solutions or B must equal zero. At $x = L$, the wavefunction must also be zero yielding the following possible values for the wavenumber, k_x.

$$k_x = \frac{n\pi}{L}, n = 1, 2, 3....$$

This analysis can now be repeated in the y and z direction. Each possible solution corresponds to a cube in k-space with size $n\pi/L$ as indicated on Figure.

$$N = 2 \times \frac{1}{8} \times \left(\frac{L}{\pi}\right)^3 \times \frac{4}{3} \times \pi \times k^3$$

A factor of two is added to account for the two possible spins of each solution. The density per unit energy is then obtained using the chain rule:

$$\frac{dN}{dE} = \frac{dN}{dk}\frac{dk}{dE} = \left(\frac{L}{\pi}\right)^3 \pi k^2 \frac{dk}{dE}$$

The kinetic energy E of a particle with mass m^* is related to the wavenumber, k, by:

$$E(k) = \frac{\hbar^2 k^2}{2m^*}, \text{ providing } \frac{dk}{dE} = \frac{m^*}{\hbar^2 k} \text{ and } k = \frac{\sqrt{2m^* E}}{\hbar}$$

And the density of states per unit volume and per unit energy, $g(E)$, becomes:

$$g(E) = \frac{1}{L^3}\frac{dN}{dE} = \frac{8\pi\sqrt{2}}{h^3} m^{*3/2} \sqrt{E}, \text{ for } E \geq 0$$

The density of states is zero at the bottom of the well as well as for negative energies.

The same analysis also applies to electrons in a semiconductor. The effective mass takes into account the effect of the periodic potential on the electron. The minimum energy of the electron is the energy at the bottom of the conduction band, E_c, so that the density of states for electrons in the conduction band is given by:

$$g_c(E) = \frac{8\pi\sqrt{2}}{h^3} m^{*3/2} \sqrt{E - E_c} \text{ for } E \geq E_c$$

$g_c(E) = 0$, for $E < E_c$

Example

Calculate the number of states per unit energy in a 100 by 100 by 10 nm piece of silicon ($m^* = 1.08\ m_0$) 100 meV above the conduction band edge. Write the result in units of eV^{-1}.

Solution

The density of states equals:

$$g(E) = \frac{8\pi\sqrt{2}}{h^3} m^{*3/2} \sqrt{E - E_c}$$

$$= \frac{8\pi\sqrt{2}\,(1.08 \times 9.1 \times 10^{-31})^{3/2}}{(6.626 \times 10^{-34})^3} \sqrt{0.1 \times 1.6 \times 10^{-19}}$$

$$= 1.51 \times 10^{56}\ m^{-3}J^{-1}$$

So that the total number of states per unit energy equals:

$$g(E)V = 1.51 \times 10^{56} \times 10^{-22}\, J^{-1} = 2.41 \times 10^5\, eV^{-1}$$

CALCULATION OF THE DENSITY OF STATES IN 1, 2 AND 3 DIMENSIONS

We will here postulate that the density of electrons in *k*–space is constant and equals the physical length of the sample divided by 2π and that for each dimension. The number of states between k and $k + dk$ in 3, 2 and 1 dimension then equals:

$$\frac{dN_{3D}}{dk} = 2(\frac{L}{2\pi})^3 4\pi k^2, \frac{dN_{2D}}{dk} = 2(\frac{L}{2\pi})^2, 2\pi k \frac{dN_{1D}}{dk} = 2(\frac{L}{2\pi})$$

We now assume that the electrons in a semiconductor are close to a band minimum, E_{min} and can be described as free particles with a constant effective mass, or:

$$E(k) = E_{\min} + \frac{h^2 k^2}{8\pi^2 m^*}$$

Elimination of k using the $E(k)$ relation above then yields the desired density of states functions, namely:

$$g_{c,3D} = \frac{dN_{3D}}{dE} = \frac{8\pi\sqrt{2}}{h^3} m^{*3/2} \sqrt{E - E_{\min}}, \text{for } E \geq E_{\min}$$

for a three-dimensional semiconductor,

$$g_{c,2D} = \frac{dN_{2D}}{dE} = \frac{4\pi m^*}{h^2}, \text{for } E \geq E_{\min}$$

For a two-dimensional semiconductor such as a quantum well in which particles are confined to a plane, and

$$g_{c,1D} = \frac{dN_{1D}}{dE} = \sqrt{\frac{2\pi m^*}{h^2}} \frac{1}{\sqrt{E - E_{\min}}}, \text{for } E \geq E_{\min}$$

For a one-dimensional semiconductor such as a quantum wire in which particles are confined along a line.

An example of the density of states in 3, 2 and 1 dimension is shown in the figure below:

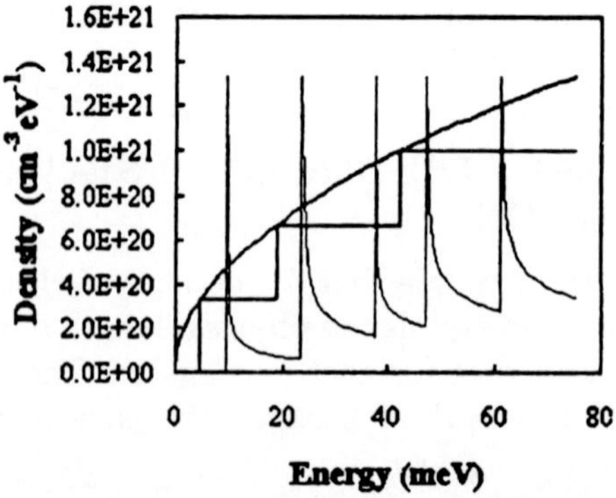

Fig. Density of states per unit volume and energy for a 3-D semiconductor (blue curve), a 10 nm quantum well with infinite barriers (red curve) and a 10 nm by 10 nm quantum wire with infinite barriers (green curve). $m^*/m_0 = 0.8$.

The above figure illustrates the added complexity of the quantum well and quantum wire: Even though the density in two dimensions is constant, the density of states for a quantum well is a step function with steps occurring at the energy of each quantized level. The case for the quantum wire is further complicated by the degeneracy of the energy levels: for instance a two-fold degeneracy increases the density of states associated with that energy level by a factor of two. A list of the degeneracy (not including spin) for the 10 lowest energies in a quantum well, a quantum wire and a quantum box, all with infinite barriers, is provided in the table below:

Quantum Well			Quantum Wire				Quantum Box				
n_x	E/E_0	Degeneracy	n_x	n_y	E/E_0	Degeneracy	n_x	n_y	n_z	E/E_0	Degeneracy
1	1	1	1	1	2	1	1	1	1	3	1
2	4	1	1	2	5	2	1	1	2	6	3
3	9	1	2	2	8	1	1	2	2	9	3
4	16	1	1	3	10	2	1	1	3	11	3
5	25	1	2	3	13	2	2	2	2	12	1
6	36	1	1	4	17	2	1	2	3	14	6
7	49	1	3	3	18	1	2	2	3	17	3
8	64	1	2	4	20	2	1	1	4	18	3
9	81	1	3	4	25	2	1	3	3	19	3
10	100	1	1	5	26	2	1	2	4	21	6
11	121	1	2	5	29	2	2	3	3	22	3
12	144	1	4	4	32	1	2	2	4	24	6
13	169	1	3	5	34	2	1	3	4	26	3
14	196	1	1	6	37	6	1	1	5	27	3
15	225	1	2	6	40	13	3	3	3	27	1
							3	2	4	29	3

Fig. Degeneracy (not including spin) of the lowest 10 energy levels in a quantum well, a quantum wire with square cross-section and a quantum cube with infinite barriers. The energy E_0 equals the lowest energy in a quantum well, which has the same size

Next, we compare the actual density of states in three dimensions with equation. While somewhat tedious, the exact number of states can be calculated as well as the maximum energy. The result is shown in Figure. The number of states in an energy range of 20 E_0 are plotted as a function of the normalized energy E/E_0. A dotted line is added to guide the eye. The solid line is calculated using equation.A clear difference can be observed between the two, while they are expected to merge for large values of E/E_0.

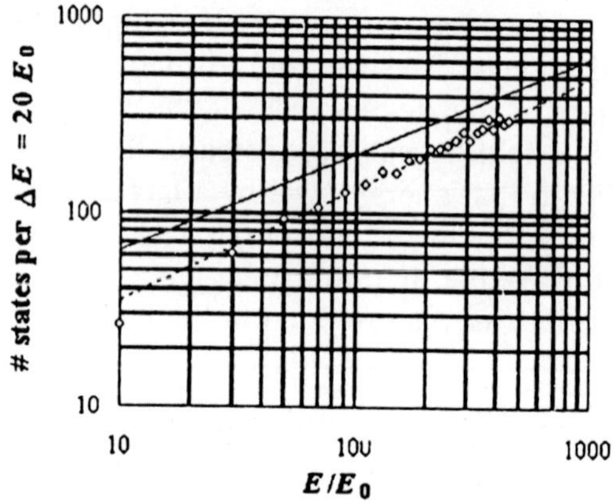

Fig. Number of states within a range $\Delta E = 20\ E_0$ as a function of the normalized energy E/E_0. (E_0 is the lowest energy in a 1-dimensional quantum well).

A comparison of the total number of states illustrates the same trend as shown in Figure. Here the solid line indicates the actual number of states, while the dotted line is obtained by integrating equation.

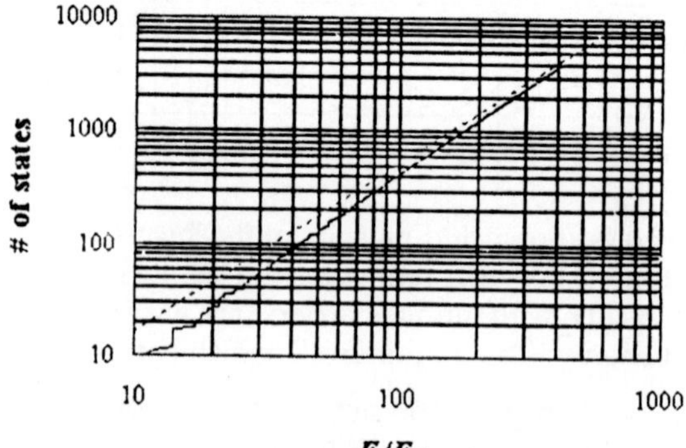

Fig. Number of states with energy less than or equal to E as a function of E_0 (E_0 is the lowest energy in an 1-dimensional quantum well). Actual number (solid line) is compared with the integral of equation (dotted line).

CARRIER DISTRIBUTION FUNCTIONS

The distribution or probability density functions describe the probability with which one can expect particles to occupy the available energy levels in a given system. Of particular interest is the probability density function of electrons, called the Fermi function. The derivation of such probability density functions belongs in a statistical thermodynamics course. However, given the importance of the Fermi distribution function, we will carefully examine an example as well as the characteristics of this function. Other distribution functions such as the impurity distribution functions, the Bose-Einstein distribution function and the Maxwell Boltzmann distribution are also provided.

FERMI-DIRAC DISTRIBUTION FUNCTION

The Fermi-Dirac distribution function, also called Fermi function, provides the probability of occupancy of energy levels by Fermions. Fermions are half-integer spin particles, which obey the Pauli exclusion principle. The Pauli exclusion principle postulates that only one Fermion can occupy a single quantum state. Therefore, as Fermions are added to an energy band, they will fill the available states in an energy band just like water fills a bucket. The states with the lowest energy are filled first, followed by the next higher ones. At absolute zero temperature (T = 0 K), the energy levels are all filled up to a maximum energy, which we call the Fermi level. No states above the Fermi level are filled. At higher temperature, one finds that the transition between completely filled states and completely empty states is gradual rather than abrupt.

Electrons are Fermions. Therefore, the Fermi function provides the probability that an energy level at energy, E, in thermal equilibrium with a large system, is occupied by an electron. The system is characterized by its temperature, T, and its Fermi energy, E_F. The Fermi function is given by:

$$f(E) = \frac{1}{1 + e^{(E - E_F)/kT}}$$

This function is plotted in Figure for different temperatures.

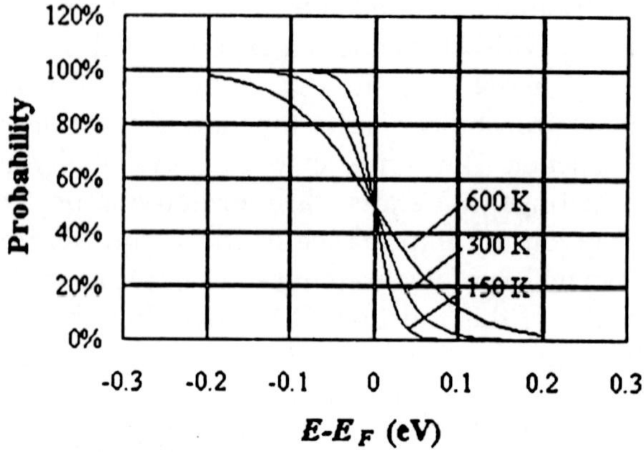

Fig. The Fermi function at three different temperatures.

The Fermi function has a value of one for energies, which are more than a few times kT below the Fermi energy. It equals 1/2 if the energy equals the Fermi energy and decreases exponentially for energies which are a few times kT larger than the Fermi energy. While at $T = 0$ K the Fermi function equals a step function, the transition is more gradual at finite temperatures and more so at higher temperatures.

EXAMPLE

To better understand the origin of distribution functions, we now consider a specific system with equidistant energy levels at 0.5, 1.5, 2.5, 3.5, 4.5, 5.5, ... eV. Each energy level can contain two electrons. Since electrons are indistinguishable from each other, no more than two electrons (with opposite spin) can occupy a given energy level. This system contains 20 electrons.

The minimum energy of this system corresponds to the situation where all 20 electrons occupy the ten lowest energy levels without placing more than 2 in any given level. This situation occurs at $T = 0$ K and the total energy equals 100 eV.

Since we are interested in a situation where the temperature is not zero, we arbitrarily set the total energy at

106 eV, which is 6 eV more than the minimum possible energy of this system. This ensures that the thermal energy is not zero so that the system must be at a non-zero temperature.

There are 24 possible and different configurations, which satisfy these particular constraints. Eight of those configurations are shown in Figure, where the filled circles represent the electrons:

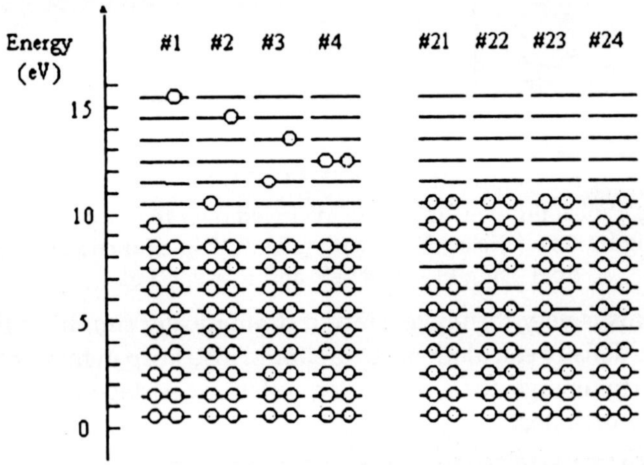

Fig. Eight of the 24 possible configurations in which 20 electrons can be placed having a total energy of 106 eV.

We now apply the basis postulate of statistical thermodynamics, namely that all possible configurations are equally likely to occur. The expected configuration therefore equals the average occupancy of all possible configurations.

The average occupancy of each energy level taken over all (and equally probable) 24 configurations is compared in Figure to the Fermi-Dirac distribution function. A best fit was obtained using a Fermi energy of 9.998 eV and kT = 1.447 eV or T = 16,800 K. The agreement is surprisingly good considering the small size of this system.

Based on the construction of the distribution function in this example, one would expect the distribution function to be dependent on the density of states. This is the case for small systems. However, for large systems and for a single energy level in thermal equilibrium with a larger system, the distribution function no longer depends on the density of states.

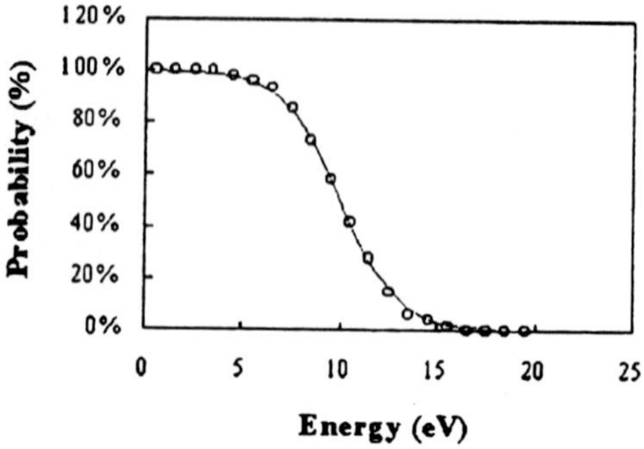

Fig. Probability versus energy averaged over the 24 possible configurations (circles) fitted with a Fermi-Dirac function (solid line) using $kT = 1.447$ eV and $E_F = 9.998$ eV.

This is very fortunate, since it dramatically simplifies the carrier density calculations. One should also keep in mind that the Fermi energy for a particular system does depend on the density of states.

IMPURITY DISTRIBUTION FUNCTIONS

The distribution function of impurities differs from the Fermi-Dirac distribution function although the particles involved are Fermions. The difference is due to the fact that an ionized donor energy level still contains one electron, which can have either spin (spin up or spin down). The donor energy level cannot be empty since this would leave a doubly positively charged atom, which would have an energy different from the donor energy. The distribution function for donors therefore differs from the Fermi function and is given by:

$$f_{donor}(E_d) = \frac{1}{1 + \frac{1}{2} e^{(E_d - E_F)/kT}}$$

The distribution function for acceptors differs also because of the different possible ways to occupy the acceptor level. The neutral acceptor contains no electrons. The ionized acceptor contains one electron, which can have either spin, while the doubly negatively charged state is not allowed since this would require a different energy. This restriction would yield a factor of 2 in front of the exponential term. In addition, one finds that most commonly used semiconductors have a two-fold degenerate valence band, which causes this factor to increase to four, yielding:

$$f_{acceptor}\left(E_\alpha\right) = \frac{1}{1 + 4e^{(E_\alpha - E_F)/kT}}$$

OTHER DISTRIBUTION FUNCTIONS AND COMPARISON

Other distribution functions include the Bose-Einstein distribution and the Maxwell-Boltzmann distribution. These are briefly discussed below and compared to the Fermi-Dirac distribution function.

The Bose-Einstein distribution function applies to bosons. Bosons are particles with integer spin and include photons, phonons and a large number of atoms. Bosons do not obey the Pauli exclusion principle so that any number can occupy one energy level. The Bose-Einstein distribution function is given by:

$$f_{BE(E)} = \frac{1}{e^{(E - E_F)/kT} - 1}$$

This function is only defined for $E > E_F$.

The Maxwell Boltzmann applies to non-interacting particles, which can be distinguished from each other. This distribution function is also called the classical distribution function since it provides the probability of occupancy for non-interacting particles at low densities. Atoms in an ideal gas form a typical example of such particles. The Maxwell-Boltzmann distribution function is given by:

$$f_{MB(E)} = \frac{1}{e^{(E - E_F)/kT}}$$

A plot of the three distribution functions, the Fermi-Dirac distribution, the Maxwell-Boltzmann distribution and the Bose-Einstein distribution is shown in Figure.

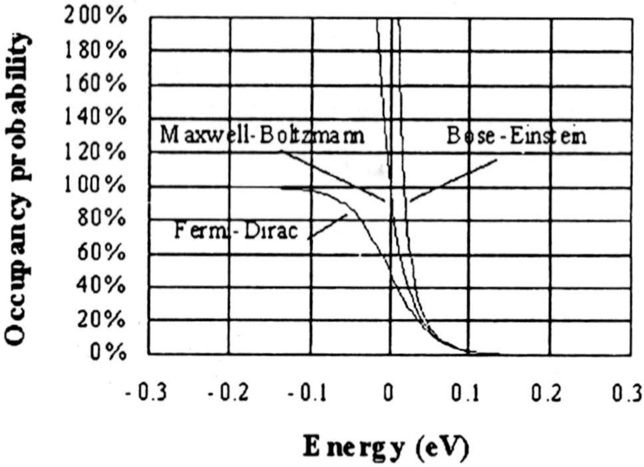

Fig. Probability of occupancy versus energy of the Fermi-Dirac, the Bose-Einstein and the Maxwell-Boltzmann distribution. The Fermi energy, E_F, is assumed to be zero.

All three functions are almost equal for large energies (more than a few kT beyond the Fermi energy). The Fermi-Dirac distribution reaches a maximum of 100per cent for energies, which are a few kT below the Fermi energy, while the Bose-Einstein distribution diverges at the Fermi energy and has no validity for energies below the Fermi energy.

DERIVATION OF THE FERMI-DIRAC DISTRIBUTION FUNCTION

To derive the Fermi-Dirac distribution function, we start from a series of possible energies, labeled E_i. At each energy, we can have g_i possible states and the number of states that are occupied equals $g_i f_i$, where f_i is the probability of occupying a state at energy E_i. We also assume that the number of possible states is very large, so that the discrete nature of the states can be ignored.

The number of possible ways - called configurations - to fit $g_i f_i$ electrons in g_i states, given the restriction that only one electron can occupy each state, equals:

$$W_i = \frac{g_i!}{(g_i - g_i f_i)! g_i f_i!}$$

This equation is obtained by numbering the individual states and exchanging the states rather than the electrons. This yields a total number of $g_i!$ possible configurations. However since the empty states are all identical, we need to divide by the number of permutations between the empty states, as all permutations cannot be distinguished from each other and can therefore only be counted once. In addition, all the filled states are indistinguishable from each other, so we need to divide also by all permutations between the filled states, namely $g_i f_i!$.

The number of possible ways to fit the electrons in the number of available states is called the multiplicity function.

The multiplicity function for the whole system is the product of the multiplicity functions for each energy E_i:

$$W = \prod_i W_i = \prod_i \frac{g_i!}{(g_i - g_i f_i)! g_i f_i!}$$

Using Stirling's approximation, one can eliminate the factorial signs, yielding:

$$\ln W = \sum_i \ln W_i = \sum_i [g_i \ln g_i - g_i(1 - f_i)\ln(g_i - g_i f_i) - g_i f_i \ln g_i f_i]$$

The total number of electrons in the system equals N and the total energy of those N electrons equals U. These system parameters are related to the number of states at each energy, g_i, and the probability of occupancy of each state, f_i, by:

$$N = \sum_i g_i f_i$$

and

$$U = \sum_i E_i g_i f_i$$

According to the basic assumption of statistical thermodynamics, all possible configurations are equally probable. The multiplicity function provides the number of configurations for a specific set of occupancy probabilities, f_i. The multiplicity function sharply peaks at the thermal equilibrium distribution since this is the most likely distribution of the system and must therefore be associated with the largest number of - equally probable - configurations. The occupancy probability in thermal equilibrium is therefore obtained by finding the maximum of the multiplicity function, W, while keeping the total energy and the number of electrons constant.

For convenience, we maximize the logarithm of the multiplicity function instead of the multiplicity function itself. According to the Lagrange method of undetermined multipliers, we must maximize the following function:

$$InW - a\sum_j g_j f_j - b\sum_j E_j g_j f_j$$

where *a* and *b* need to be determined. The maximum of the multiplicity function, W, is obtained from:

$$\frac{\partial}{\partial(g_i f_i)}\left[InW - a\sum_j g_j f_i - a\sum_j E_j g_j f_i\right] = 0$$

which can be solved, yielding:

$$In\frac{g_i - g_i f_i}{g_i f_i} - a - bE_i = 0$$

or

$$f_i = fFD(E_i) = \frac{1}{1 + \exp(a + bE_i)}$$

which can be written in the following form

$$fFD(E_i) = \frac{1}{1 + \exp\left(\frac{{}_i - E_F}{\beta}\right)}$$

with $\beta = 1/b$ and $E_F = -a/b$. The symbol E_F was chosen since this constant has units of energy and will be the constant associated with this probability distribution.

Taking the derivative of the total energy, one obtains:

$$dU = \sum_i E_i d(g_i f_i) + \sum_i g_i f_i dE_i$$

Using the Lagrange equation, this can be rewritten as:

$$dU = \beta d(InW) + \sum_i g_i f_i dE_i + E_F dN$$

Any variation of the energies, E_I, can only be caused by a change in volume, so that the middle term can be linked to a volume variation dV.

$$dU = \beta d(InW) + \left[\sum_i g_i f_i \frac{dE_i}{dV}\right] dV + E_F dN$$

Comparing this to the thermodynamic identity:

$$dU = TdS - pdV + \mu dN$$

one finds that $\beta = kT$ and $S = k \ln W$, where k is a constant that must be determined . The energy, E_F, equals the energy associated with the particles, namely the electro-chemical potential, μ.

The comparison also identifies the entropy, S, as being proportional to the logarithm of the multiplicity function, W. The proportionality constant, k, is known as Boltzmann's constant.

The Fermi-Dirac distribution function then becomes:

$$f_{FD}(E) = \frac{1}{1 + \exp\left(\frac{E - E_F}{kT}\right)}$$

Note that this derivation can only truly be followed if one has prior knowledge of statistical thermodynamics. Those who are well versed in this field can quickly derive the Fermi-Dirac and other distribution functions using the Gibbs sum.

CARRIER DENSITIES

Now that we have discussed the density of states and the distribution functions, we have all the necessary tools to calculate the carrier density in a semiconductor.

GENERAL DISCUSSION

The density of electrons in a semiconductor is related to the density of available states and the probability that each of these states is occupied. The density of occupied states per unit volume and energy, $n(E)$,), is simply the product of the density of states in the conduction band, $g_c(E)$ and the Fermi-Dirac probability function, $f(E)$, (also called the Fermi function):

$$n(E) = g_c(E) f(E)$$

Since holes correspond to empty states in the valence band, the probability of having a hole equals the probability that a particular state is not filled, so that the hole density per unit energy, $p(E)$, equals:

$$p(E) = g_v(E)\left[1 - f(E)\right]$$

Where $g_v(E)$ is the density of states in the valence band. The density of carriers is then obtained by integrating the density of carriers per unit energy over all possible energies within a band. A general expression is derived as well as an approximate analytic solution, which is valid for non-degenerate semiconductors. In addition, we also present the Joyce-Dixon approximation, an approximate solution useful when describing degenerate semiconductors.

The density of states in a semiconductor was obtained by solving the Schrödinger equation for the particles in the semiconductor. Rather than using the actual and very complex potential in the semiconductor, we use the simple particle-in-a box model, where one assumes that the particle is free to move within the material.

For an electron which behaves as a free particle with effective mass, m^*, the density of states yielding:

$$g_c(E) = \frac{8\pi\sqrt{2}}{h^3} m_e^{*3/2} \sqrt{E - E_c}, \text{ for } E \geq E_c$$

where E_c is the bottom of the conduction band below which the density of states is zero. The density of states for holes in the valence band is given by:

$$g_v(E) = \frac{8\pi\sqrt{2}}{h^3} m_h^{*3/2} \sqrt{E_v - E}, \text{ for } E \leq E_v$$

CALCULATION OF THE FERMI INTEGRAL.

The carrier density in a semiconductor, is obtained by integrating the product of the density of states and the probability density function over all possible states.

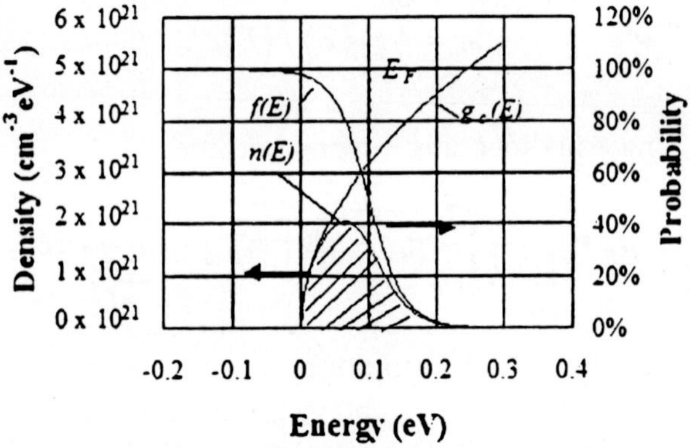

Fig. The carrier density integral. Shown are the density of states, $g_c(E)$, the density per unit energy, $n(E)$, and the probability of occupancy, $f(E)$. The carrier density, n_o, equals the crosshatched area.

For electrons in the conduction band the integral is taken from the bottom of the conduction band, labeled, E_c, to the top of the conduction band:

$$n = \int_{E_c}^{\text{top of the conduction band}} n(E)dE = \int_{E_c}^{\text{top of the conduction band}} g_c(E)f(E)dE$$

Where $g_c(E)$ is the density of states in the conduction band and $f(E)$ is the Fermi function.

This general expression is illustrated with Figure for a parabolic density of states function with $E_c = 0$. The figure shows the density of states function, $g_c(E)$, the Fermi function, $f(E)$, as well as the product of both, which is the density of electrons per unit volume and per unit energy, $n(E)$. The integral corresponds to the crosshatched area.

The actual location of the top of the conduction band does not need to be known as the Fermi function goes to zero at higher energies. The upper limit can therefore be replaced by infinity. We also relabeled the carrier density as n_0 to indicate that the carrier density is the carrier density in thermal equilibrium.

$$n_0 = \int_{E_c}^{\infty} g_c(E) f(E) dE$$

Using equations this integral becomes:

$$n_o = \int_{E_c}^{\infty} \frac{8\pi\sqrt{2}}{h^3} m_e^{*3/2} \sqrt{E - E_c} \frac{1}{1 + e^{\frac{E - E_F}{kT}}} dE$$

While this integral can not be solved analytically at non-zero temperatures, we can obtain either a numeric solution or an approximate analytical solution. Similarly for holes one obtains:

$$p_o = \int_{-\infty}^{E_v} g_v(E)[1 - f(E)] dE$$

and

$$p_o = \int_{-\infty}^{E_v} \frac{8\pi\sqrt{2}}{h^3} m_h^{*3/2} \sqrt{E_v - E} \frac{1}{1 + e^{\frac{E_F - E}{kT}}} dE$$

The calculation of the electron and hole density in a semiconductor is further illustrated by Figure.

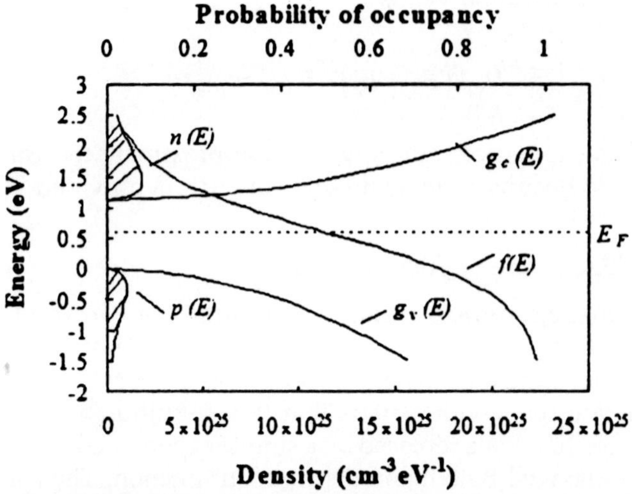

Fig. The density of states and carrier densities in the conduction and valence band. Shown are the electron and hole density per unit energy, $n(E)$ and $p(E)$, the density of states in the conduction and valence band, $g_c(E)$ and $g_v(E)$ and the probability of occupancy, $f(E)$. The crosshatched area indicates the electron and hole densities.

Indicated on the figure are the density of states in the conduction and valence band, the Fermi distribution function and the electron and hole densities per unit energy. The crosshatched areas indicate the thermal-equilibrium carrier densities. From the figure, one can easily see that the electron density will increase as the Fermi energy is increased. The hole density decreases with increasing Fermi energy. As the Fermi energy is decreased, the electron density decreases and the hole density increases.

Carrier Density at Zero Kelvin

Equation can be solved analytically at $T = 0$ K, since the Fermi function at $T = 0$ K equals one for all energies below the Fermi energy and 0 for all energies larger than the Fermi energy. Equation can therefore be simplified to:

$$n_o = \int_{E_c}^{E_F} g_c(E)dE \text{ at } T = 0\,\text{K}$$

and integration yields:

$$n_0 = \frac{2}{3}\frac{\sqrt{2}}{\pi^2}\frac{m_e^{*3/2}}{\hbar^3}(E_F - E_c)^{3/2} \text{ for } E_F \geq E_c$$

This expression can be used to approximate the carrier density in heavily degenerate semiconductors provided $kT << (E_F - E_c) > 0$

Non-degenerate Semiconductors

Non-degenerate semiconductors are defined as semiconductors for which the Fermi energy is at least $3kT$ away from either band edge. The reason we restrict ourselves to non-degenerate semiconductors is that this definition allows the Fermi function to be replaced by a simple exponential function, i.e. the Maxwell-Boltzmann distribution function. The carrier density integral can then be solved analytically yielding:

$$p_0 = \frac{2}{3}\frac{\sqrt{2}}{\pi^2}\frac{m_h^{*3/2}}{\hbar^2}(E_v - E_F)^{3/2} \text{ for } E_v \geq E_F$$

with

$$n_0 \cong \int_{E_c}^{\infty} \frac{8\pi\sqrt{2}}{h^3} m_e^{*3/2} \sqrt{E - E_c} e \frac{E_F - E}{kT} dE = N_c e^{\frac{E_F - E_c}{kT}}$$

where N_c is the *effective density of states* in the *conduction band*. The Fermi energy, E_F, is obtained from:

Similarly for holes, one can approximate the hole density integral as:

with

$$p_o \cong \int_{-\infty}^{E_v} \frac{8\pi\sqrt{2}}{h^3} m_h^{*3/2} \sqrt{E_v - E} e^{\frac{E - E_F}{kT}} dE = N_v e^{\frac{E_v - E_F}{kT}}$$

where N_v is the effective density of states in the *valence* band. The Fermi energy, E_F, is obtained from:

Example: Calculate the effective densities of states in the conduction and valence bands of germanium, silicon and gallium arsenide at 300 K.

Solution

The effective density of states in the conduction band of germanium equals:

$$N_c = 2\left(\frac{2\pi m_e^* kT}{h^2}\right)^{3/2}$$

$$= 2\left(\frac{2\pi 0.55\times 9.11\times 10^{-31}\times 1.38\times 10^{-23}\times 300}{(6.626\times 10^{-34})^2}\right)^{3/2}$$

$$= 1.02\times 10^{25}\ \text{m}^{-3} = 1.02\times 10^{19}\ \text{cm}^{-3}$$

where the effective mass for density of states was used. Similarly one finds the effective densities for silicon and gallium arsenide and those of the valence band:

	Germanium	**Silicon**	**Gallium Arsenide**
Nc (cm^{-3})	1.02×10^{19}	2.81×10^{19}	4.35×10^{17}
Nv (cm^{-3})	5.64×10^{18}	1.83×10^{19}	7.57×10^{18}

Note that the effective density of states is temperature dependent and can be obtain from:

$$N_c(T) = N_c(300\text{K})\left(\frac{T}{300}\right)^{3/2}$$

where N_c(300 K) is the effective density of states at 300 K.

Degenerate semiconductors

A useful approximate expression applicable to degenerate semiconductors was obtained by Joyce and Dixon and is given by:

for electrons and by:

$$\frac{E_F - E_c}{kT} \cong \ln\frac{n_o}{N_c} + \frac{1}{\sqrt{8}}\frac{n_o}{N_c} - \left(\frac{3}{16} - \frac{\sqrt{3}}{9}\right)\left(\frac{n_o}{N_c}\right)^2 + \ldots$$

for holes.

INTRINSIC SEMICONDUCTORS

Intrinsic semiconductors are semiconductors, which do not contain impurities. They do contain electrons as well as holes. The electron density equals the hole density since the thermal activation of an electron from the valence band to the conduction band yields a free electron in the conduction band as well as a free hole in the valence band. We will identify the intrinsic hole and electron density using the symbol n_i, and refer to it as the intrinsic carrier density.

Intrinsic Carrier Density

Intrinsic semiconductors are usually non-degenerate, so that the expressions for the electron and hole densities in non-degenerate semiconductors apply. Labeling the Fermi energy of intrinsic material as E_i, we can then write two relations between the intrinsic carrier density and the intrinsic Fermi energy, namely:

$$\frac{E_v - E_F}{kT} \cong \ln\frac{p_o}{N_v} + \frac{1}{\sqrt{8}}\frac{p_o}{N_v} - \left(\frac{3}{16} - \frac{\sqrt{3}}{9}\right)\left(\frac{p_o}{N_v}\right)^2 + \ldots$$

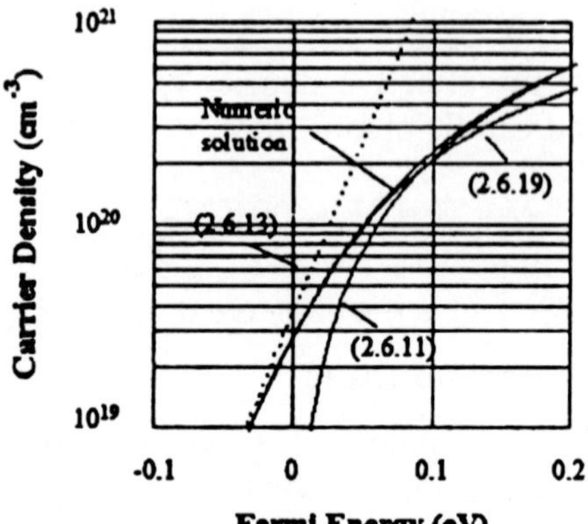

Fig. Intrinsic carrier density versus temperature in gallium arsenide (GaAs), silicon (Si) and germanium (Ge).

It is possible to eliminate the intrinsic Fermi energy from both equations, simply by multiplying both equations and taking the square root. This provides an expression for the intrinsic carrier density as a function of the effective density of states in the conduction and valence band, and the bandgap energy $E_g = E_c - E_v$.

$$n_i = n_o|_{(E_F = E_i)} = N_c e^{(E_i - E_c)/kT}$$

$$n_i = p_o|_{(E_F = E_i)} = N_v e^{(E_v - E_i)/kT}$$

The temperature dependence of the intrinsic carrier density is dominated by the exponential dependence on the energy bandgap. In addition, one has to consider the temperature dependence of the effective densities of states and that of the energy bandgap. A plot of the intrinsic carrier density versus temperature is shown in Figure. The temperature dependence of the effective masses was ignored.

Example

Calculate the intrinsic carrier density in germanium, silicon and gallium arsenide at 300, 400, 500 and 600 K.

Solution

The intrinsic carrier density in silicon at 300 K equals:

$$n_i(300\text{K}) = \sqrt{N_c N_v}\exp(\frac{-E_g}{2kT})$$

$$= \sqrt{2.81\times 10^{19}\times 1.83\times 10^{19}}\exp(\frac{-1.12}{2\times 0.0258})$$

$$= 8.72\times 10^{9}\ \text{cm}^{-3}$$

Similarly one finds the intrinsic carrier density for germanium and gallium arsenide at different temperatures, yielding:

	Germanium	Silicon	Gallium Arsenide
300K	2.02×10^{13}	8.72×10^{9}	2.03×10^{6}
400K	1.38×10^{15}	4.52×10^{12}	5.98×10^{9}
500K	1.91×10^{16}	2.16×10^{14}	7.98×10^{11}
600K	1.18×10^{17}	3.07×10^{15}	2.22×10^{13}

Note that the values at 300 K as calculated in example. This is due to an accumulation of assumptions in the derivation.

Mass Action Law

One can prove that the product of the electron and hole density equals the square of the intrinsic carrier density for any non-degenerate semiconductor. By multiplying the expressions for the electron and hole densities in a non-degenerate semiconductor, as in equations one obtains:

This property is referred to as the *mass action law*. It is a powerful relation, which enabies to quickly find the hole density if the electron density is known or vice versa. This relation is only valid for non-degenerate semiconductors in thermal equilibrium

Intrinsic Fermi energy

The above equations for the intrinsic electron and hole density can be solved for the intrinsic Fermi energy, yielding:

The intrinsic Fermi energy is typically close to the *midgap* energy, half way between the conduction and valence band edge. The intrinsic Fermi energy can also be expressed as a function of the effective masses of the electrons and holes in the semiconductor. For this we use equations for the effective density of states in the conduction and valence band, yielding:

Intrinsic Material as Reference

Dividing the expressions for the carrier densities, by the one for the intrinsic density allows to write the carrier densities as a function of the intrinsic density and the intrinsic Fermi energy, E_i, or:

$$E_i = \frac{E_c + E_v}{2} + \frac{3}{4} kT \ln\left(\frac{m_h^*}{m_e^*}\right)$$

We will use primarily these two equations to find the electron and hole density in a semiconductor in thermal equilibrium. The same relations can also be rewritten to obtain the Fermi energy from either carrier density, namely:

DOPED SEMICONDUCTORS

Doped semiconductors are semiconductors, which contain impurities, foreign atoms incorporated into the crystal structure of the semiconductor. Either these impurities can be unintentional, due to lack of control during the growth of the semiconductor, or they can be added on purpose to provide free carriers in the semiconductor.

The generation of free carriers requires not only that impurities are present, but also that the impurities give off electrons to the conduction band in which case they are called *donors*. If they give off holes to the valence band, they are called *acceptors* (since they effectively accept an electron from the filled valence band). The ionization of shallow donors and acceptors are illustrated by Figure. Indicated are the donor and acceptor energies, E_d and E_a. The donor energy level is filled prior to ionization. Ionization causes the donor to be emptied, yielding an electron in the conduction band and a positively charged donor ion. The acceptor energy is empty prior to ionization. Ionization of the acceptor corresponds to the empty acceptor level being filled by an electron from the filled valence band. This is equivalent to a hole given off by the acceptor atom to the valence band.

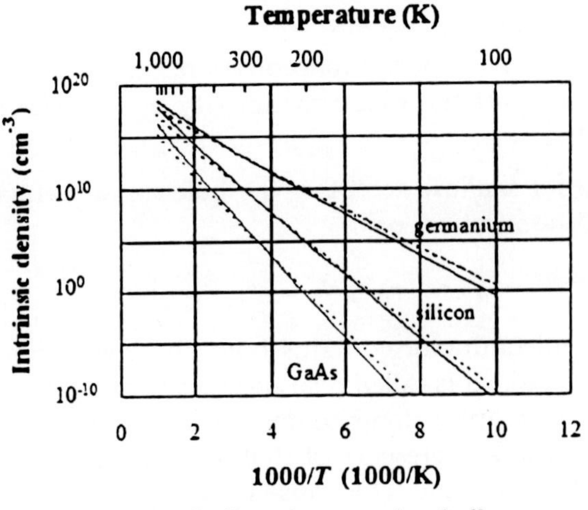

Fig. Ionization of a shallow donor and a shallow acceptor

A semiconductor doped with impurities, which are ionized (meaning that the impurity atoms either have donated or accepted an electron), will therefore contain free carriers. *Shallow impurities* are impurities, which require little energy - typically around the thermal energy, kT, or less - to ionize. *Deep impurities* require energies much larger than the thermal energy to ionize so that only a fraction of the impurities present in the semiconductor contribute to free carriers. Deep impurities, which are more than five times the thermal energy away from either band edge, are very unlikely to ionize. Such impurities can be effective recombination centres, in which electrons and holes fall and annihilate each other. Such deep impurities are also called *traps*.

Ionized donors provide free electrons in a semiconductor, which is then called *n-type*, while ionized acceptors provide free holes in a semiconductor, which we refer to as being a *p-type* semiconductor.

Dopants and Impurities

The ionization of the impurities is dependent on the thermal energy and the position of the impurity level within the energy band gap as described by the impurity distribution functions.

Shallow impurities readily ionize so that the free carrier density equals the impurity concentration. For shallow donors this implies that the electron density equals the donor concentration, or:

While for shallow acceptors the hole density equals the acceptor concentration, or:

$$n_0 \cong N_d^+ = N_d$$

If a semiconductor contains both shallow donors and shallow acceptors it is called *compensated* since equal amounts of donor and acceptor atoms compensate each other, yielding no free carriers. The presence of shallow donors and shallow acceptors in a semiconductor cause the electrons given off by the donor atoms to fall into the acceptor state, which ionizes the acceptor atoms without yielding a free electron or hole. The resulting carrier density in compensated material, which contains both shallow donors and shallow acceptors, is approximately equal to the difference between the donor and acceptor concentration if the donor concentration is larger, yielding n-type material, or:

If the acceptor concentration is larger than the donor concentration, the hole density of the resulting p-type material equals the difference between the acceptor and donor concentration, or:

Ionization Energy Model

The energy required to remove an electron from a donor atom can be approximated using a hydrogen-like model. After all, the donor atom consists of a positively charged ion and an electron just like the proton and electron of the hydrogen atom. The difference however is that the average distance, *r*, between the electron and the donor ion is much larger since the electron occupies one of the outer orbitals. This is illustrated by Figure.

Trajectory of an electron bound to a donor ion within a semiconductor crystal.

For shallow donors, this distance, *r*, is much larger than the inter-atomic spacing of the semiconductor crystal. The ionization energy, E_d, can be estimated by modifying equation, which describes the electron energy in a Hydrogen atom, yielding:

where m^*_{cond} is the effective mass for conductivity calculations and ε_r is the relative dielectric constant of the semiconductor. The ionization energy is calculated as the difference between the energy of a free electron and that of an electron occupying the lowest energy level, E_1.

Example

Calculate the ionization energy for shallow donors and acceptors in germanium and silicon using the hydrogen-like model.

Solution

Using the effective mass for conductivity calculations one finds the ionization energy for shallow donors in germanium to be:

$$E_c - E_d = 13.6 \frac{m^*_{cond}}{m_0 \varepsilon_r^2} \text{ eV} = 13.6 \frac{0.12}{16^2} \text{ eV} = 6.4 \text{ meV}$$

The calculated ionization energies for donors and acceptors in germanium and silicon are provided below.

	Germanium	Silicon
donors	6.4meV	13.8meV
acceptors	11.2meV	20.5meV

Note that the actual ionization energies differ from these values and depend on the actual donor atom.

Analysis of Non-degenerately Doped Semiconductors

The calculation of the electron density starts by assuming that the semiconductor is neutral, so that there is a zero charge density in the material. This is a reasonable assumption since a net charge density would result in an electric field. This electric field would move any mobile charge so that it eliminates any charge imbalance.

The charge density in a semiconductor depends on the free electron and hole density and on the ionized impurity densities. Ionized donors, which have given off an electron, are positively charged. Ionized acceptors, which have accepted an electron, are negatively charged. The total charge density is therefore given by:

$$E_c - E_d = 13.6 \frac{m^*_{cond}}{m_0 \varepsilon_r^2} \text{ eV}$$

The hole concentration in thermal equilibrium can be written as a function of the electron density by using the mass action law. This yields the following relation between the electron density and the ionized impurity densities:

$$\rho = q(p_o - n_o + N_d^+ - N_a^-) = 0$$

Note that the use of the mass action law restricts the validity of this derivation to non-degenerate semiconductors as defined in section. Solving this quadratic equation yields a solution for the electron density, namely:

The same derivation can be repeated for holes, yielding:

The above expressions provide the free carrier densities for compensated semiconductors assuming that all donors and acceptors are ionized.

From the carrier densities, one then obtains the Fermi energies using equations which are repeated below:

or

The Fermi energies in n-type and p-type silicon as a function of doping density is shown in Figure for different temperatures:

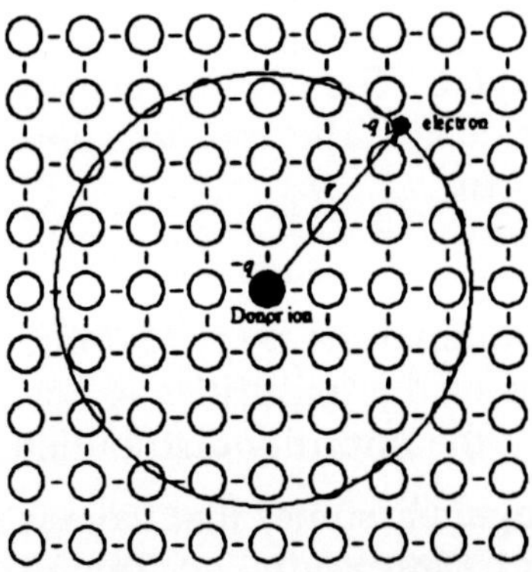

Fig. Fermi energy of n-type and p-type silicon, $E_{F,n}$ and $E_{F,p}$, as a function of doping density at 100, 200, 300, 400 and 500 K. Shown are the conduction and valence band edges, E_c and E_v. The midgap energy is set to zero.

Figure illustrates how the Fermi energies vary with doping density. The Fermi energy varies linearly, when plotting the density on a logarithmic scale, up to a doping density of 10^{18} cm^{-3}. This simple dependence requires that the semiconductor is neither intrinsic nor degenerate and that all the dopants are ionized. For compensated material, containing only shallow dopants, one uses the net doping density, $|N_d - N_a|$.

Example

A germanium wafer is doped with a shallow donor density of $3n_i/2$. Calculate the electron and hole density.

Solution

The electron density is obtained from equation and the hole density is obtained using the mass action law:

Example

A silicon wafer is doped with a shallow acceptor doping of 10^{16} cm^{-3}. Calculate the electron and hole density.

Solution

Since the acceptor doping is much larger than the intrinsic density and much smaller than the effective density of states, the hole density equals: The electron density is then obtained using the mass action law The approach described in example yields the same result.

General Analysis

A more general analysis takes also into account the fact that the ionization of the impurities is not 100per cent, but instead is given by the impurity distribution functions.

The analysis again assumes that there is no net charge in the semiconductor (*charge neutrality*). This also means that the total density of positively charged particles (holes and ionized donors) must equals the total density of negatively charged particles (electrons and ionized acceptors) yielding:

$$p_o = \frac{N_a^- - N_d^+}{2} + \sqrt{\left(\frac{N_a^- - N_d^+}{2}\right)^2 + n_i^2}$$

The electron and hole densities are then written as a function of the Fermi energy. For non-degenerate semiconductors one uses equations, while the ionized impurity densities equal the impurity density multiplied with the probability of occupancy for the acceptors and one minus the probability of occupancy for the donors. The Joyce-Dixon approximation, is used to calculate the degenerate carrier densities.

A graphical solution to equation above can be obtained by plotting both sides of the equation as a function of the Fermi energy as illustrated in Figure.

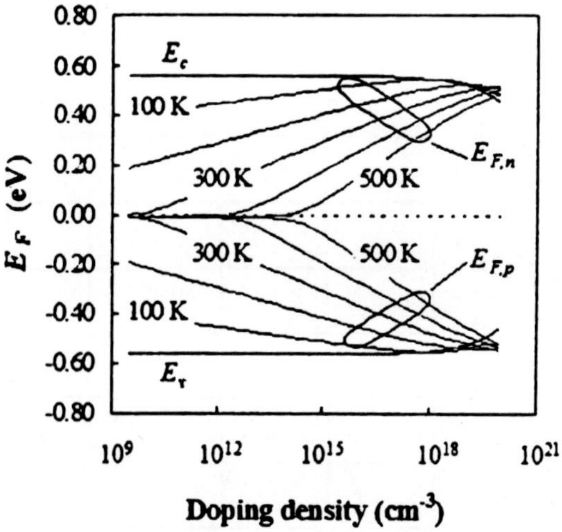

Fig. Graphical solution of the Fermi energy based on the general analysis. The value of the Fermi energy and the free carrier density is obtained at the intersection of the two curves, which represent the total positive and total negative charge in the semiconductor. N_a equals 10^{16} cm^{-3} and N_d equals 10^{14} cm^{-3}.

Figure shows the positive and negative charge densities as well as the electron and hole densities as a function of the Fermi energy. The dotted lines indicate the position of the acceptor and donor energies. The Fermi energy is obtained at the intersection of both curves as indicated by the arrow.

This graphical solution is a very useful tool to explore the Fermi energy as a function of the doping densities, ionization energies and temperature.

Operation of devices over a wide temperature range requires a detailed knowledge of the carrier density as a function of temperature. At intermediate temperatures the carrier density approximately equals the net doping, $|N_a - N_d|$. Semiconductors, which satisfy this condition, are also called extrinsic semiconductors. The free carrier density increases at high temperatures for which the intrinsic density approaches the net doping density and decreases at low temperatures due to incomplete ionization of the dopants. The carrier density and Fermi energy are shown in Figure for silicon doped with 10^{16} cm^{-3} donors and 10^{15} cm^{-3} acceptors:

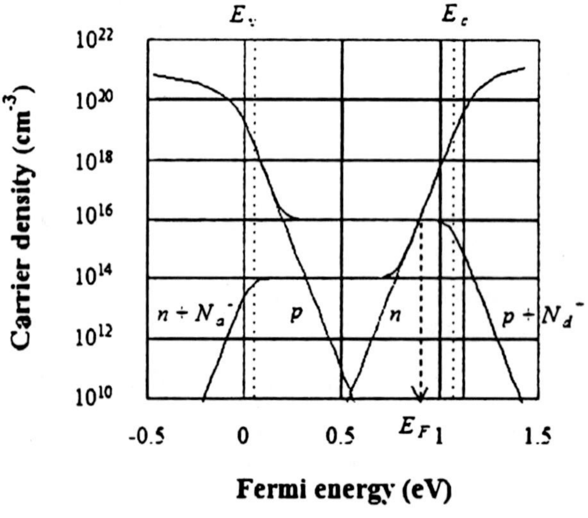

Fig. Electron density and Fermi energy as a function of temperature in silicon with $N_d = 10^{16}$ cm^{-3}, $N_a = 10^{14}$ cm^{-3} and $E_c - E_d = E_a - E_v = 50$ meV. The activation energy at 70 K equals 27.4 meV.

At high temperatures, the carrier density equals the intrinsic carrier concentration, while at low temperatures the carrier density is dominated by the ionization of the donors.

The temperature dependence is related to an activation energy by fitting the carrier density versus $1/T$ on a semi-logarithmic scale to a straight line of the form $n_o(T) = C$ exp(-

E_A/kT), where C is a constant. At high temperatures this activation energy equals half the bandgap energy or $E_A = E_g/2$.

The temperature dependence at low temperatures is somewhat more complex as it depends on whether or not the material is compensated. Figure was calculated for silicon containing both donors and acceptors. At 70 K the electron density is below the donor density but still larger than the acceptor density. Under such conditions the activation energy, E_A, equals half of the ionization energy of the donors or $(E_c - E_d)/2$. At lower temperatures where the electron density is lower than the acceptor density, the activation energy equals the ionization energy or $E_c - E_d$. This behaviour is explained by the fact that the Fermi energy in compensated material is fixed at the donor energy. The donors levels are always partially empty as electrons are removed from the donor atoms to fill the acceptor energy levels. If the acceptor density is smaller than the electron density – as is true for uncompensated material - the Fermi energy does change with temperature and the activation energy approaches half of the ionization energy.

Lightly doped semiconductors suffer from freeze-out at relatively high temperature. Higher-doped semiconductors freeze-out at lower temperatures. Highly-doped semiconductors do not contain a single donor energy, but rather an impurity band which overlaps with the conduction or valence band. The overlap of the two bands results in free carriers even at zero Kelvin. Degenerately doping a semiconductor therefore eliminates freeze-out effects.

Temperature Dependence of the Carrier Density in Non-degenerate Material

The analysis of non-degenerate semiconductors can be further expanded to include carrier freeze-out in the presence of donors as well as acceptors. Equation can be written as:

For n-type material ($N_d > N_a$), this equation can be simplified to:

$$p_o + \frac{N_d}{1 + 2\exp\frac{E_F - E_d}{kT}} = n_o + \frac{N_a}{1 + 4\exp\frac{E_a - E_F}{kT}}$$

where the hole density, p_o, has been ignored and the acceptors are assumed to be fully ionized since the Fermi energy is expected to be above the energy of even the deepest acceptors.

Example

4H-SiC is doped with 2×10^{17} cm^{-3} nitrogen donor atoms ($E_c - E_d = 90$ meV). Use $N_c = 4 \times 10^{20}$ cm^{-3}.

a. Calculate the electron density at 300 K.

b. Calculate the hole density at 300 K after adding 2×10^{18} cm^{-3} aluminum acceptor atoms (E_a - E_c = 220 meV). Use $N_v = 1.6 \times 10^{20}$ cm^{-3}.

Solution

First we calculate N^*

$$N^* = \frac{N_c}{2}\exp\frac{E_d - E_c}{kT} = 6.16 \times 10^{17}\ \text{cm}^{-3}$$

The free electron density is then obtained from:

$$n_o = -\frac{N^*}{2} + \sqrt{\frac{(N^*)^2}{4} + N^* N_d} = 4.3 \times 10^{16}\ \text{cm}^{-3}$$

As a result 21.5 per cent of the donors are ionized

a. Since we are now dealing with p-type material we have to recalculate N^*

$$N^{**} = \frac{N_v}{4}\exp\frac{E_v - E_a}{kT} = 8.08 \times 10^{16}\ \text{cm}^{-3}$$

where the factor 4 is due to the doubly degenerate valence band. The free hole density is then obtained from:

$$p_o = -\frac{N^* + N_d}{2} + \sqrt{\frac{(N^* + N_d)^2}{4} + N^*(N_a - N_d)}$$
$$= 5.52 \times 10^{16}\ \text{cm}^{-3}$$

which is identical to except that the donor and acceptor densities have been exchanged. Only 2.76 per cent of the acceptors are ionized while all donors are ionized.

NON-EQUILIBRIUM CARRIER DENSITIES

Up until now, we have only considered the thermal equilibrium carrier densities, n_0 and p_0. However most devices of interest are not in thermal equilibrium. Keep in mind that a constant ambient constant temperature is not a sufficient condition for thermal equilibrium. In fact, applying a non-zero voltage to a device or illuminating it with light will cause a non-equilibrium condition, even if the temperature is constant.

To describe a system that is not in thermal equilibrium we assume that each of the carrier distributions is still in equilibrium with itself. Such assumption is justified on the basis that electrons readily interact with each other and interact with holes only on a much longer time scale. As a result the electron density can still be calculated using the Fermi-Dirac distribution function, but with a different value for the Fermi energy. The total carrier density for a non-degenerate semiconductor is then described by:

$$n_o = -\frac{N^{\bullet} + N_a}{2} + \sqrt{\frac{(N^{\bullet} + N_a)^2}{4} + N^{\bullet}(N_d - N_a)}$$

Where δ_n is the *excess electron density* and F_n is the *quasi-Fermi energy* for the electrons. Similarly, the hole density can be expressed as:

Where δp is the *excess hole density* and F_p is the *quasi-Fermi energy* for the holes.

Example A piece of germanium doped with 10^{16} cm^{-3} shallow donors is illuminated with light generating 10^{15} cm^{-3} excess electrons and holes. Calculate the quasi-Fermi energies relative to the intrinsic energy and compare it to the Fermi energy in the absence of illumination.

Solution

The carrier densities when illuminating the semiconductor are:

$$n = n_o + \delta n = 10^{16} + 10^{15} = 1.1 \times 10^{16} \text{ cm}^{-3}$$

$$p = p_o + \delta p \cong 10^{15} \text{ cm}^{-3}$$

and the quasi-Fermi energies are:

$$F_n - E_i = kT \ln \frac{n}{n_i} = 0.0259 \times \ln \frac{1.1 \times 10^{16}}{2 \times 10^{13}} = 163 \text{meV}$$

$$F_p - E_i = -kT \ln \frac{p}{n_i} = 0.0259 \times \ln \frac{1 \times 10^{15}}{2 \times 10^{13}} = -101 \text{meV}$$

For comparison, the Fermi energy in the absence of light equals

$$E_F - E_i = kT \ln \frac{n_o}{n_i} = 0.0259 \times \ln \frac{10^{16}}{2 \times 10^{13}} = 161 \text{meV}$$

which is very close to the quasi-Fermi energy of the majority carriers.

CARRIER TRANSPORT

Any motion of free carriers in a semiconductor leads to a current. This motion can be caused by an electric field due to an externally applied voltage, since the carriers are charged particles. We will refer to this transport mechanism as carrier drift. In addition, carriers also move from regions where the carrier density is high to regions where the carrier density is low. This carrier transport mechanism is due to the thermal energy and the associated random motion of the carriers. We will refer to this transport mechanism as carrier diffusion. The total current in a semiconductor equals the sum of the drift and the diffusion current.

As one applies an electric field to a semiconductor, the electrostatic force causes the carriers to first accelerate and then reach a constant average velocity, v, due to collisions with impurities and lattice vibrations. The ratio of the velocity to the applied field is called the mobility. The velocity saturates at high electric fields reaching the saturation velocity. Additional scattering occurs when carriers flow at the surface of a semiconductor, resulting in a lower mobility due to surface or interface scattering mechanisms.

Diffusion of carriers is obtained by creating a carrier density gradient. Such gradient can be obtained by varying the doping density in a semiconductor or by applying a thermal gradient.

Both carrier transport mechanisms are related since the same particles and scattering mechanisms are involved. This leads to a relationship between the mobility and the diffusion constant called the Einstein relation.

CARRIER DRIFT

The motion of a carrier drifting in a semiconductor due to an applied electric field, ε, is illustrated in Figure. The field causes the carrier to move with a velocity, *v*.

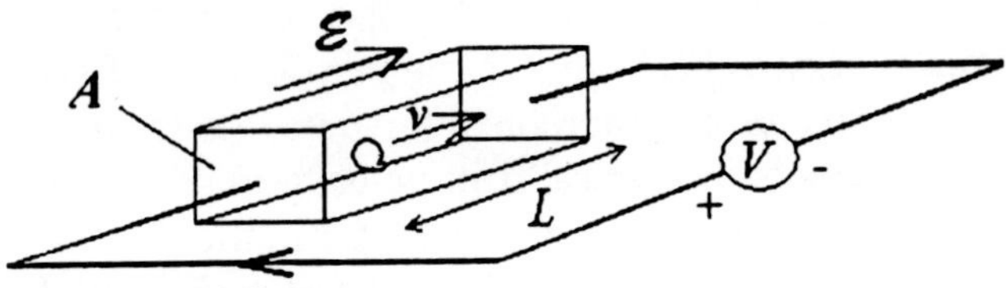

Fig. Drift of a carrier due to an applied electric field.

Assuming that all the carriers in the semiconductor move with the same velocity, the current can be expressed as the total charge in the semiconductor divided by the time needed to travel from one electrode to the other, or:

$$I = \frac{Q}{t_r} = \frac{Q}{L/v}$$

where τ_r is the transit time of a particle, traveling with velocity, *v*, over the distance *L*. The current density can then be rewritten as a function of the charge density, ρ:

$$\vec{J} = \frac{Q}{AL}\vec{v} = \rho\vec{v} = qn\vec{v}$$

If the carriers are negatively charged electrons, the current density equals:

$$\vec{J} = -qn\vec{v}$$

while for positively charged holes it is:

$$\vec{J} = qp\vec{v}$$

Where n and p are the electron and hole density in the semiconductor.

It should be noted that carriers do not follow a straight path along the electric field lines. Instead they bounce around in the semiconductor and constantly change direction and velocity due to scattering. This behaviour occurs even when no electric field is applied and is due to the thermal energy of the carriers. Thermodynamics teaches us that electrons in a non-degenerate and non-relativistic electron gas have a thermal energy of $kT/2$ per particle per degree of freedom. A typical thermal velocity at room temperature is around 10^7 cm/s, which exceeds the typical drift velocity in semiconductors. The carrier motion in the semiconductor in the absence and in the presence of an electric field can therefore be visualized as in Figure.

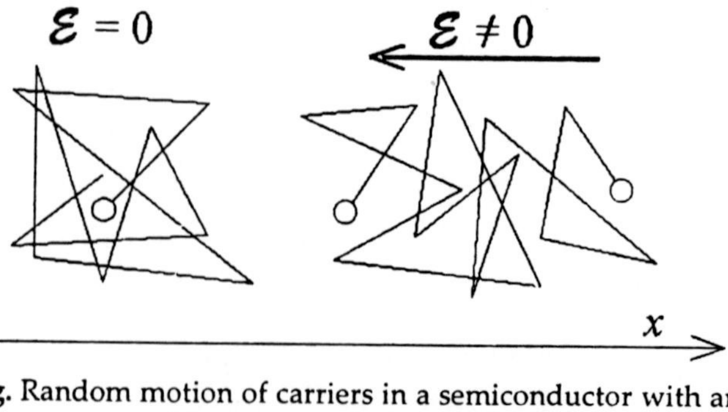

Fig. Random motion of carriers in a semiconductor with and without an applied electric field.

In the absence of an applied electric field, the carrier exhibits random motion and the carriers move quickly through the semiconductor and frequently change direction. When an electric field is applied, the random motion still occurs but in addition, there is on average a net motion along the direction of the field. Due to their different electronic charge, holes move on average in the direction of the applied field, while electrons move in the opposite direction.

We now analyze the carrier motion considering only the average velocity, $\langle \vec{v} \rangle$ of the carriers. Applying Newton's law, we state that the acceleration of the carriers is proportional to the applied force:

$$\vec{F} = m\vec{a} = m\frac{d < \vec{v} >}{dt}$$

The force consists of the difference between the electrostatic force and the scattering force due to the loss of momentum at the time of scattering. This scattering force equals the momentum divided by the average time between scattering events, so that:

$$\vec{F} = q\vec{\mathcal{E}} - \frac{m < \vec{v} >}{\tau_c}$$

where a particle with charge q was assumed.

Combining both relations yields an expression for the average particle velocity:

$$q\vec{\mathcal{E}} = m\frac{d < \vec{v} >}{dt} + \frac{m < \vec{v} >}{\tau_c}$$

We now consider only the steady state situation in which the particle has already accelerated and has reached a constant average velocity. Under such conditions, the velocity is proportional to the applied electric field and we define the mobility as the velocity to field ratio:

$$\mu \stackrel{\Delta}{=} \frac{|\vec{v}|}{|\vec{\mathcal{E}}|} = \frac{q\tau_c}{m}$$

The mobility of a particle in a semiconductor is therefore expected to be large if its mass is small and the time between scattering events is large.

The drift current, described by can then be rewritten as a function of the mobility, yielding:

$$\vec{J} = qn\mu_n \vec{\mathcal{E}}$$

For electrons and

$$\vec{J} = qp\mu_p \vec{\mathcal{E}}$$

For holes

Throughout this derivation, we simply considered the mass, m, of the particle. However in order to incorporate the effect of the periodic potential of the atoms in the semiconductor we must use the effective mass, m^*, rather than the free particle mass:

$$\mu = \frac{q\tau_c}{m^*}$$

Example

Electrons in undoped gallium arsenide have a mobility of 8,800 cm^2/V-s. Calculate the average time between collisions. Calculate the distance traveled between two collisions (also called the mean free path). Use an average velocity of 10^7 cm/s.

SolutionThe collision time, τ_c, is obtained from:

$$\tau_c = \frac{\mu_n m_e^*}{q} = \frac{0.88 \times 0.067 \times 9.1 \times 10^{-31}}{1.6 \times 10^{-19}} = 0.34 \text{ ps}$$

where the mobility was first converted in MKS units.

The mean free path, l, equals:

$$l = v_{average}\tau_c = 10^7 \times 0.34 \times 10^{-12} = 34 \text{ nm}$$

Impurity Scattering

Impurities are foreign atoms in the solid, which are efficient scattering centers especially when they have a net charge. Ionized donors and acceptors in a semiconductor are a common example of such impurities. The amount of scattering due to electrostatic forces between the carrier and the ionized impurity depends on the interaction time and the number of impurities. Larger impurity concentrations result in a lower mobility. The dependence on the interaction time helps to explain the temperature dependence. The interaction time is directly linked to the relative velocity of the carrier and the impurity, which is related to the thermal velocity of the carriers. The thermal velocity increases with the ambient temperature so that the interaction time decreases. Thereby, the amount of scattering decreases, resulting in a mobility increase with temperature. To first order, the mobility due to impurity scattering is proportional to $T^{3/2}/N_I$, where N_I is the density of charged impurities.

Lattice Scattering

Scattering by lattice waves includes the absorption or emission of either acoustical or optical phonons. These phonons represent quanta of mechanical waves that travel through the semiconductor crystal. Since the density of phonons in a solid increases with temperature, the scattering time due to this mechanism will decrease with temperature as will the mobility. Theoretical calculations reveal that the mobility in non-polar semiconductors, such as silicon and germanium, is dominated by *acoustic* phonon interaction. The resulting mobility is expected to be proportional to $T^{-3/2}$, while the mobility due to *optical* phonon scattering only is expected to be proportional to $T^{-1/2}$. Experimental values of the temperature dependence of the mobility.in germanium, silicon and gallium arsenide are provided in Table.

	Germanium	Silicon	Gallium Arsenide
Electron mobility	aT–1.7	aT–2.4aT–1.0	
Hole mobility	aT–2.3	aT–2.2aT–2.1	

Temperature dependence of the mobility in germanium, silicon and gallium arsenide due to phonon scattering

Surface Scattering

The surface and interface mobility of carriers is affected by the nature of the adjacent layer or surface. Even if the carrier does not transfer into the adjacent region, its wavefunction does extend over 1 to 10 nanometer, so that there is a non-zero probability that the particle is in the adjacent region. The net mobility is then a combination of the mobility in both layers. For carriers in the inversion layer of a MOSFET, one finds that the mobility can be up to three times lower than the bulk value. This is due to the distinctly lower mobility of electrons in the amorphous silicon oxide. The presence of charged surface states further reduces the mobility just as ionized impurities would.

CARRIER MOBILITY

Doping dependence

The mobility of electrons and holes in silicon at room temperature is shown in Figure.

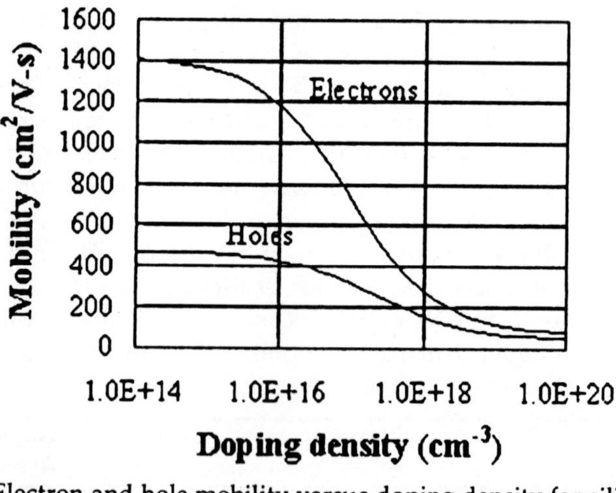

Fig. Electron and hole mobility versus doping density for silicon.

The electron mobility and hole mobility have a similar doping dependence: For low doping concentrations, the mobility is almost constant and is primarily limited by phonon scattering. At higher doping concentrations, the mobility decreases due to ionized impurity scattering with the ionized doping atoms. The actual mobility also depends on the type of dopant. Figure is for phosphorous and boron doped silicon.

Note that the mobility is linked to the total number of ionized impurities or the *sum* of the donor and acceptor densities. The free carrier density, is to first order related to the *difference* between the donor and acceptor concentration.

The minority carrier mobility also depends on the total impurity density. The minority-carrier mobility can be approximated by the majority-carrier mobility in a material with the same number of impurities.

The mobility at a particular doping density is obtained from the following empiric expression:

$$\mu = \mu_{min} + \frac{\mu_{max} - \mu_{min}}{1 + \left(\frac{N}{N_r}\right)^{\alpha}}$$

where μ_{min}, μ_{max}, α and N_r are fit parameters. These parameters for arsenic, phosphorous and boron doped silicon are provided in Table.

Table:

	Arsenic	**Phosphorous**	**Boron**
μmin(cm²/V-s)	52.2	68.5	44.9
μmax(cm²/V-s)	1417	1414	470.5
N_r(cm⁻³)	9.68 × 106	9.20 × 1016	2.23 × 1017
α	0.68	0.711	0.719

Parameters for calcuiation of the mobility as a function of the doping density

The resulting mobilities in units of cm^2/V-s are listed for different doping densities in Table.

Mobility in silicon for different dopants and doping densities.

N	Arsenic	Phosphorous	Boron
10^{15} cm^{-3}	1359	1362	462
10^{16} cm^{-3}	1177	1184	429
10^{17} cm^{-3}	727	721	317
10^{18} cm^{-3}	284	277	153
10^{19} cm^{-3}	108	115	71

Conductivity and Resistivity

The conductivity of a material is defined as the current density divided by the applied electric field. Since the current density equals the product of the charge of the mobile carriers, their density and velocity as described by equations it can be expressed as a function of the electric field using the mobility. To include the contribution of electrons as well as holes to the conductivity, we add the current density due to holes to that of the electrons, or:

$$J = qnv_e + qpv_n = q(n\mu_n + p\mu_p)\varepsilon$$

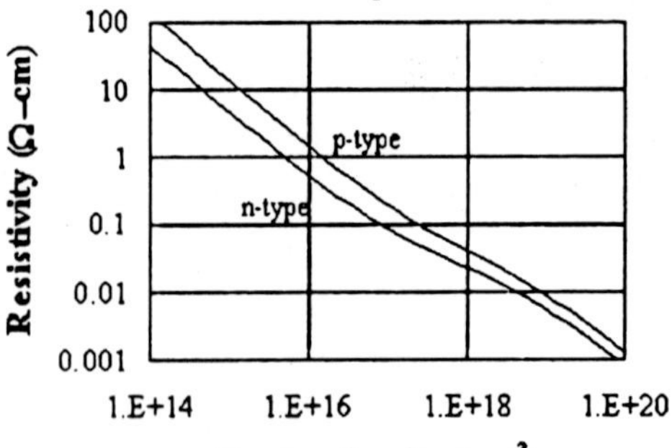

Fig. Resistivity of n-type and p-type silicon versus doping density.

The conductivity due to electrons and holes is then obtained from:

$$\sigma = \frac{\Delta J}{\varepsilon} = q\left(n\mu_n + p\mu_p\right)$$

The resistivity is defined as the inverse of the conductivity, namely:

$$\rho = \frac{1}{\sigma} = \frac{1}{q\left(\mu_n n + \mu_p p\right)}$$

The resulting resistivity as calculated with the expression above is shown in Figure.

Sheet Resistance

The sheet resistance concept is used to characterize both wafers and thin doped layers, since it is typically easier to measure the sheet resistance rather than the resistivity of the material. The sheet resistance of a uniformly-doped layer with resistivity, ρ, and thickness, *t*, is given by their ratio:

$$R_S = \frac{\rho}{t}$$

While the unit of the sheet resistance is Ohms, one refers to it as Ohms per square. This nomenclature comes in handy when the resistance of a rectangular piece of material with length, *L*, and width *W* must be obtained. It equals the product of the sheet resistance and the number of squares or:

$$R = R_S \frac{L}{W}$$

where the number of squares equals the length divided by the width. Figure provides, as an example, the sheet resistance of a 14 mil thick silicon wafer for both *n*-type and *p*-type silicon.

Example

A piece of silicon doped with arsenic ($N_d = 10^{17}$ cm^{-3}) is 100 μm long, 10 μm wide and 1 μm thick. Calculate the resistance of this sample when contacted one each end.

SolutionThe resistivity of the silicon equals:

$$\rho = \frac{1}{qn\mu_n} = \frac{1}{1.6 \times 10^{-19} \times 10^{17} \times 727} = 0.086 \Omega\text{cm}$$

where the mobility was obtained from Table.

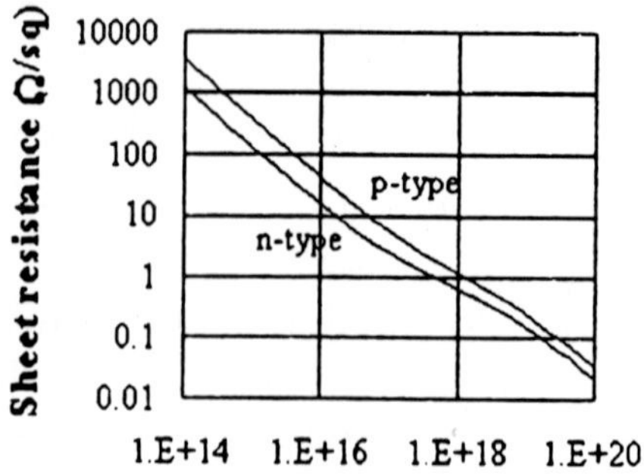

Doping density (cm^{-3})

Fig. Sheet resistance of a 14 mil thick n-type and p-type silicon wafer versus doping density.

The resistance then equals:

$$R = \rho \frac{L}{Wt} = 0.086 \times \frac{100 \times 10^{-4}}{10 \times 10^{-4} \times 10^{-4}} = 8.6 k\Omega$$

An alternate approach is to first calculate the sheet resistance, R_s:

$$R_s = \frac{\rho}{t} = \frac{0.086}{10^{-4}} = 860 \Omega / \text{square}$$

From which one then obtains the resistance:

$$R = R_s \frac{L}{W} = 860 \times \frac{100 \times 10^{-4}}{10 \times 10^{-4}} = 8.6 k\Omega$$

VELOCITY SATURATION

The linear relationship between the average carrier velocity and the applied field breaks down when high fields are applied. As the electric field is increased, the average carrier velocity and the average carrier energy increases as well. When the carrier energy increases beyond the optical phonon energy, the probability of emitting an optical phonon increases abruptly. This mechanism causes the carrier velocity to saturate with increasing electric field. For carriers in silicon and other materials, which do not contain accessible higher bands, the velocity versus field relation increases monotonically as shown in Figure.

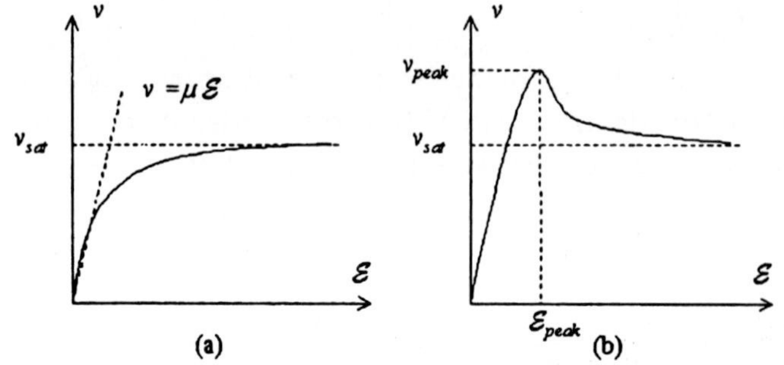

Fig. Velocity-field relation for a) materials without accessible higher bands and b) materials with an accessible higher band

For those materials the velocity-field relation can be described by:

$$v(\varepsilon) = \frac{\mu\varepsilon}{1 + \frac{\mu\varepsilon}{v_{sat}}}$$

The maximum obtainable velocity, v_{sat}, is referred to as the saturation velocity.

The analysis is more complex for materials such as GaAs and InP, which contain multiple closely spaced conduction band minima. Electrons originating in the lowest Γ minimum accelerate under high field conditions and gain enough energy to transfer to the X or L minima. Since the density of states in the X and L minima is much higher because of the higher effective mass, the probability that the electron transfers back into the Γ minimum is small because of the small number of available states. The electrons therefore remain in the higher minima as long as the electric field exceeds the peak field. The corresponding velocity initially increases with increasing applied electric field. However as carriers transfer to the higher minimum with a higher effective mass, the average carrier velocity decreases. This causes a distinct peak in the velocity field curve as shown in Figure. Such velocity-field curves are therefore characterized by the peak velocity, v_{peak}/i>, *and the peak field,* E_{peak}, *in addition to the mobility,* μ, *and the saturation velocity,* v_{sat}.

A first estimate of the saturation velocity, v_{sat}, of a material can be obtained by calculating the carrier velocity when it reaches an energy equal to the optical phonon energy.

$$v_{sat} = \sqrt{\frac{2E_{phonon}}{m^*}}$$

While this equation can at best provide a rough estimate of the actual saturation velocity, it does indicate the right trend. Materials with a small effective mass and high optical phonon energy are more likely to have a high saturation velocity. Materials with multiple band minima can have a rather low saturation velocity, relative to the peak velocity, if the carriers in the higher minima have a larger effective mass.

CARRIER DIFFUSION

Carrier diffusion is due to the thermal energy, *kT*, which causes the carriers to move at random even when no field is applied. This random motion does not yield a net flow of carriers nor does it yield a net current in material with a uniform carrier density as any carrier which leaves a specific location is on average replace by another one. However if a carrier gradient is present, the diffusion process will even out the carrier density variations: carriers *diffuse* from regions where the density is high to regions where the density is low. The diffusion process is not unlike the motion of sand on a vibrating table; hills as well as valleys are smoothed out over time.

In this section we will first derive the expression for the current due to diffusion and then combine it with the drift current to obtain the total drift-diffusion current.

Diffusion Current

The derivation is based on the basic notion that carriers at non-zero temperature (Kelvin) have an additional thermal energy, which equals $kT/2$ per degree of freedom. It is the thermal energy, which drives the diffusion process. At $T = 0$ K there is no diffusion.

The reader should recognize that the random nature of the thermal energy would normally require a statistical treatment of the carriers. Instead we will use average values to describe the process. Such approach is justified on the basis that a more elaborate statistical approach yields the same results. To further simplify the derivation, we will derive the diffusion current for a one-dimensional semiconductor in which carriers can only move along one direction.

We now introduce the average values of the variables of interest, namely the thermal velocity, v_{th}, the collision time, τ_c, and the mean free path, l. The thermal velocity is the average velocity of the carriers going in the positive or negative direction. The collision time is the time during which carriers will move with the same velocity before a collision occurs with an atom or with another carrier. The mean free path is the average length a carrier will travel between collisions. These three averages are related by:

$$v_{th} = \frac{l}{r}$$

Consider now the situation illustrated with Figure.

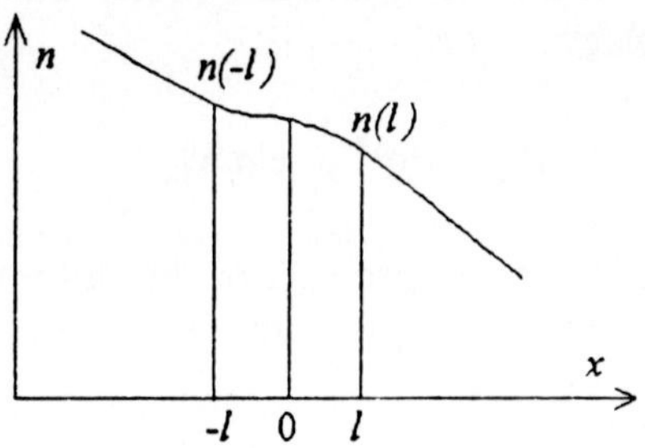

Fig. Carrier density profile used to derive the diffusion current expression

Shown is a variable carrier density, $n(x)$. Of interest are the carrier densities which are one mean free path away from $x = 0$, since the carriers, which will arrive at $x = 0$ originate either at $x = -l$ or $x = l$. The flux at $x = 0$ due to carriers that originate at $x = -l$ and move from left to right equals:

$$\Phi_{n,left \to right} = \frac{1}{2} v_{th} n(x = -l)$$

where the factor 1/2 is due to the fact that only half of the carriers move to the left while the other half moves to the right. The flux at $x = 0$ due to carriers that originate at $x = +l$ and move from right to left, equals:

$$\Phi_{n,right \to left} = \frac{1}{2} v_{th} n(x = l)$$

The total flux of carriers moving from left to right at $x = 0$ therefore equals:

$$\Phi_n = \Phi_{n,left \to right} - \Phi_{n,right \to left} = \frac{1}{2} v_{th} [n(x = -l) - n(x = l)]$$

Where the flux due to carriers moving from right to left is subtracted from the flux due to carriers moving from left to right. Given that the mean free path is small we can write the difference in densities divided by the distance between $x = -l$ and $x = l$ as the derivative of the carrier density:

$$\Phi_n = -l v_{th} \frac{n(x = l) - n(x = -l)}{2l} = -l v_{th} \frac{dn}{dx}$$

The electron diffusion current equals this flux times the charge of an electron, or:

$$J_n = -q\Phi_n = q l v_{th} \frac{dn}{dx}$$

We now replace the product of the thermal velocity and the mean free path by a single parameter, namely the diffusion constant, D_n.

$$J_n = q D_n \frac{dn}{dx}$$

Repeating the same derivation for holes yields:

$$J_p = -qD_p \frac{dp}{dx}$$

We now further explore the relation between the diffusion constant and the mobility. At first, it seems that there should be no relation between the two since the driving force is distinctly different: diffusion is caused by thermal energy while an externally applied field causes drift. However one essential parameter in the analysis, namely the collision time, τ_c, should be independent of what causes the carrier motion.

We now combine the relation between the velocity, mean free path and collision time,

$$v_{th} = \frac{l}{r}$$

with the result from thermodynamics, stating that electrons carry a thermal energy which equals $kT/2$ for each degree of freedom. Applied to a one-dimensional situation, this leads to:

$$\frac{kT}{2} = \frac{m^* v_{th}^2}{2}$$

We now use these relations to rewrite the product of the thermal velocity and the mean free path as a function of the carrier mobility:

$$lv_{th} = \frac{m^* v_{th}^2}{q} \frac{qr}{m^*} = \frac{kT}{q} \mu$$

Using the definition of the diffusion constant we then obtain the following expressions which are often referred to as the Einstein relations:

$$D_n = \mu_n \frac{kT}{q} = \mu_n V_t$$

$$D_p = \mu_n \frac{kT}{q} = \mu_p V_t$$

Example

The hole density in an n-type silicon wafer ($N_d = 10^{17}$ cm^{-3}) decreases linearly from 10^{14} cm^{-3} to 10^{13} cm^{-3} between $x = 0$ and $x = 1$ μm. Calculate the hole diffusion current density.

SolutionThe hole diffusion current density equals:

$$J_p = qD_p \frac{dp}{dx} = 1.6 \times 10^{-19} \times 8.2 \times \frac{9 \times 10^{13}}{10^{-4}} = 1.18 \text{ A/cm}^2$$

where the diffusion constant was calculated using the Einstein relation:

$$D_p = V_t \mu_p = 0.0259 \times 317 = 8.2 cm^2 / s$$

and the hole mobility in the n-type wafer was obtained from Table as the hole mobility in a p-type material with the same doping density.

Total Current

The total electron current is obtained by adding the current due to diffusion to the drift current, resulting in:

$$J_n = qn\mu_n \varepsilon + qD_n \frac{dn}{dx}$$

and similarly for holes:

$$J_p = qp\mu_p \mathcal{E} - qD_p \frac{dp}{dx}$$

The total current is the sum of the electron and hole current densities multiplied with the area, *A*, perpendicular to the direction of the carrier flow:

THE HALL EFFECT

The Hall effect describes the behaviour of the free carriers in a semiconductor when applying an electric as well as a magnetic field. The experimental setup shown in Figure, depicts a semiconductor bar with a rectangular cross section and length *L*. A voltage V_x is applied between the two contacts, resulting in a field along the x-direction. The magnetic field is applied in the z-direction.

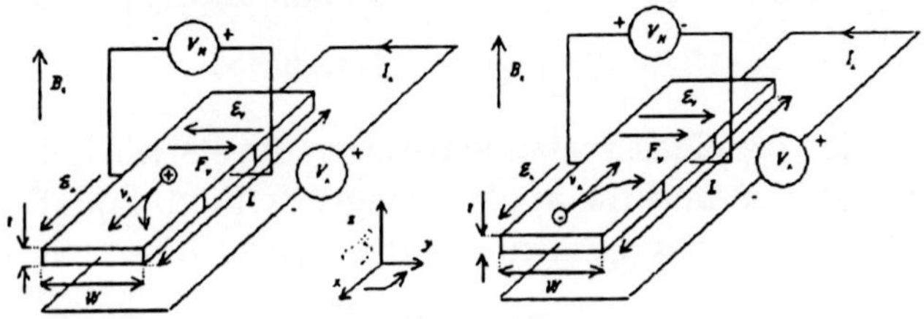

Fig. Hall setup and carrier motion for a) holes and b) electrons.

As shown in Figure the holes move in the positive x-direction. The magnetic field causes a force to act on the mobile particles in a direction dictated by the right hand rule. As a result there is a force, F_y, along the positive y-direction, which moves the holes to the right. In steady state this force is balanced by an electric field, ε_y, so that there is no net force on the holes. As a result there is a voltage across the sample, which can be measured with a high-impedance voltmeter. This voltage, V_H, is called the Hall voltage. For the sign convention shown in, the Hall voltage is positive for holes.

The behaviour of electrons is shown in figure. The electrons travel in the negative x-direction. Therefore the force, F_y, is in the positive y-direction due to the negative charge and the electrons move to the right, just like holes. The balancing electric field, ε_y, now has the opposite sign, which results in a negative Hall voltage.

To calculate the Hall field, we first calculate the Lorentz force acting on the free carriers:

$$\vec{F} = q\left(\vec{\varepsilon} \times \vec{v} \times \vec{B}\right)$$

We now assume that the carriers can only flow along the x-direction and label their velocity vx. The Lorentz force then becomes:

$$\vec{F} = q\vec{\mathcal{E}} + q\begin{vmatrix} \vec{e}_x & \vec{e}_y & \vec{e}_z \\ v_x & 0 & 0 \\ 0 & 0 & B_z \end{vmatrix} = q\mathcal{E}_x\,\vec{e}_x + q(\mathcal{E}_y - v_x B_z)\,\vec{e}_y + q\mathcal{E}_z\,\vec{e}_z$$

Since the carriers only flow along the *x*-direction, the net force must be zero along the *y* and *z* direction. As a result, the electric field is zero along the *z* direction and:

$$F_y = q\left(\varepsilon_y - v_x B_x\right) = 0$$

which provides a relation between the electric field along the *y*-direction and the applied magnetic field. Equation can also be rewritten as a function of the current density, J_x, using, yielding:

$$\varepsilon_y = v_x B_x = \frac{J_x}{qp_p0} B_z$$

This electric field is called the Hall field. The Hall coefficient, R_H, is defined as the Hall field divided by the current density, J_x, and magnetic field, B_z:

$$R_H \triangleq \frac{\varepsilon_y}{J_x B_z} = \frac{1}{qp_{p0}}$$

Once the Hall coefficient is obtained one easily finds the hole density:

$$p_{p0} = \frac{1}{qR_H} = \frac{J_x B_x}{q\varepsilon_y}$$

The carrier mobility can also be extracted from the Hall coefficient:

$$\mu_p = \frac{J_x}{qp_{p0}\varepsilon_x} = \frac{R_H}{\rho}$$

Where ρ is the resistivity of the semiconductor.

Alternatively, one can calculate the Hall coefficient from the measured current, I_x, and measured voltage, V_H:

$$R_H = \frac{V_H}{I_x B_z}\frac{tL}{W}$$

A measurement of the Hall voltage is often used to determine the type of semiconductor (*n*-type or *p*-type) the free carrier density and the carrier mobility. Repeating the measurement at different temperatures allows one to measure the free carrier density as well as the mobility as a function of temperature. Since the measurement can be done on a small piece of uniformly doped material it is by far the easiest measurement to determine the carrier mobility. It should be noted that the scattering mechanisms in the presence of a magnetic field are different and that the measured Hall mobility can differ somewhat from the drift mobility. A measurement of the carrier density versus temperature provides information regarding the ionization energies of the donors and acceptor that are present in the semiconductor. While the interpretation of the Hall measurement is straightforward in the case of a single dopant, multiple types of impurities and the presence of electrons and holes can make the interpretation non-trivial.

CARRIER RECOMBINATION AND GENERATION

Recombination of electrons and holes is a process by which both carriers annihilate each other: electrons occupy - through one or multiple steps - the empty state associated with a hole. Both carriers eventually disappear in the process. The energy difference between the initial and final state of the electron is released in the process. This leads to one possible classification of the recombination processes. In the case of radiative recombination, this energy is emitted in the form of a photon. In the case of non-radiative recombination, it is passed on to one or more phonons and in the case of Auger recombination it is given off in the form of kinetic energy to another electron. Another classification scheme considers the individual energy levels and particles involved. These different processes are further illustrated with Figure.

Band-to-band recombination occurs when an electron moves from its conduction band state into the empty valence band state associated with the hole. This band-to-band transition is typically also a radiative transition in direct bandgap semiconductors.

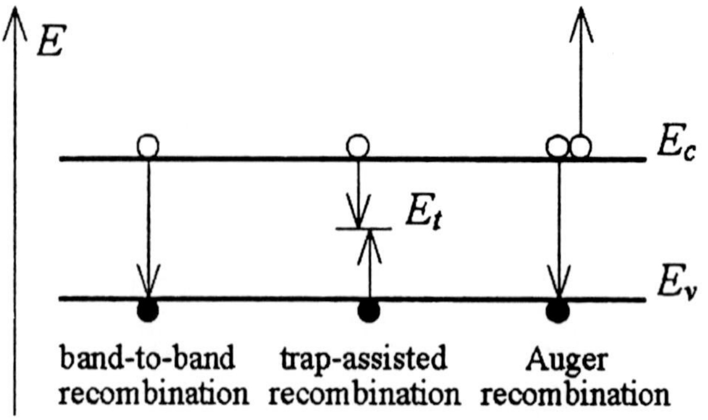

Fig. Carrier recombination mechanisms in semiconductors

Trap-assisted recombination occurs when an electron falls into a "trap", an energy level within the bandgap caused by the presence of a foreign atom or a structural defect. Once the trap is filled it cannot accept another electron. The electron occupying the trap, in a second step, moves into an empty valence band state, thereby completing the recombination process. One can envision this process as a two-step transition of an electron from the conduction band to the valence band or as the annihilation of the electron and hole, which meet each other in the trap. We will refer to this process as Shockley-Read-Hall (SRH) recombination.

Auger recombination is a process in which an electron and a hole recombine in a band-to-band transition, but now the resulting energy is given off to another electron or hole. The involvement of a third particle affects the recombination rate so that we need to treat Auger recombination differently from band-to-band recombination.

Each of these recombination mechanisms can be reversed leading to carrier generation rather than recombination. A single expression will be used to describe recombination as well as generation for each of the above mechanisms.

In addition, there are generation mechanisms, which do not have an associated recombination mechanism, such as generation of carriers by light absorption or by a high-energy electron/particle beam. These processes are referred to as ionization processes. Impact ionization, which is the generation mechanism associated with Auger recombination, also belongs to this category. The generation mechanisms are illustrated with Figure 2.8.2.

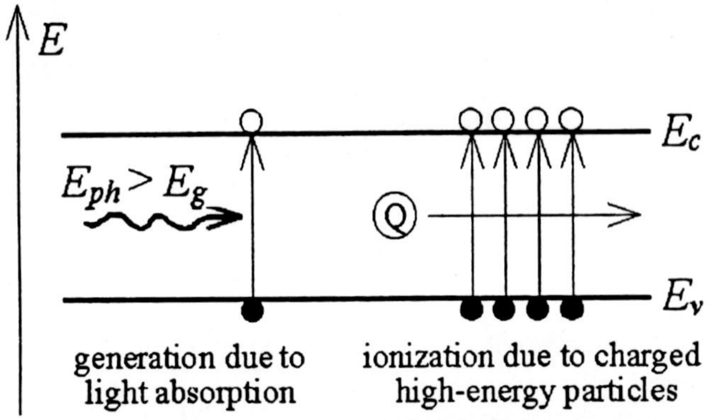

Fig. Carrier generation due to light absorption and ionization due to high-energy particle beams

Carrier generation due to light absorption occurs if the photon energy is large enough to raise an electron from the valence band into an empty conduction band state, thereby generating one electron-hole pair. The photon energy needs to be larger than the bandgap energy to satisfy this condition. The photon is absorbed in this process and the excess energy, $E_{ph} - E_g$, is added to the electron and the hole in the form of kinetic energy.

Carrier generation or ionization due to a high-energy beam consisting of *charged* particles is similar except that the available energy can be much iarger than the bandgap energy so that multiple electron-hole pairs can be formed. The high-energy particle gradually loses its energy and eventually stops. This generation mechanism is used in semiconductor-based nuclear particle counters. As the number of ionized electron-hole pairs varies with the energy of the particle, one can also use such detector to measure the particle energy.

Finally, there is a generation process called *impact ionization,* the generation mechanism that is the counterpart of Auger recombination. Impact ionization is caused by an electron/hole with an energy, which is much larger/smaller than the conduction/valence band edge. The detailed mechanism is illustrated with Figure.

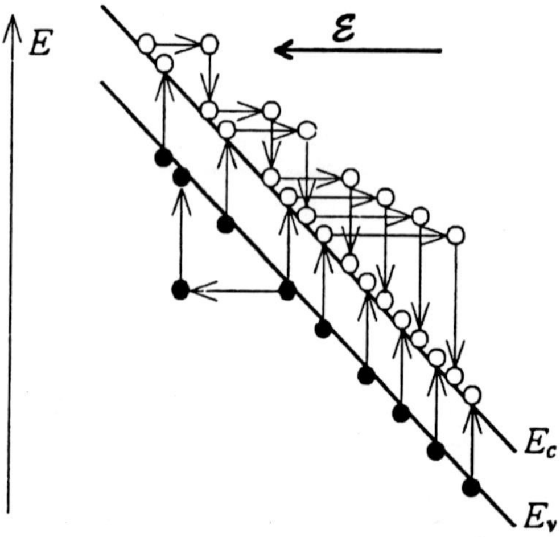

Fig. Impact ionization and avalanche multiplication of electrons and holes in the presence of a large electric field.

The excess energy is given off to generate an electron-hole pair through a band-to-band transition. This generation process causes avalanche multiplication in semiconductor diodes under high reverse bias: As one carrier accelerates in the electric field it gains energy. The kinetic energy is given off to an electron in the valence band, thereby creating an electron-hole pair. The resulting two electrons can create two more electrons which generate four more causing an avalanche multiplication effect. Electrons as well as holes contribute to avalanche multiplication.

SIMPLE RECOMBINATION-GENERATION MODEL

A simple model for the recombination-generation mechanisms states that the recombination-generation rate is proportional to the excess carrier density. It acknowledges the fact that no net recombination takes place if the carrier density equals the thermal equilibrium value. The resulting expression for the recombination of electrons in a p-type semiconductor is given by:

$$U_n = R_n - G_n = \frac{n_p - n_{p0}}{\tau_n}$$

and similarly for holes in an n-type semiconductor:

$$U_p = R_p - G_p = \frac{p_n - p_{n0}}{\tau_p}$$

where the parameter τ can be interpreted as the average time after which an excess minority carrier recombines.

We will show for each of the different recombination mechanisms that the recombination rate can be simplified to this form when applied to minority carriers in a "quasi-neutral" semiconductor. The above expressions are therefore only valid under these conditions. The recombination rates of the majority carriers equals that of the minority carriers since in steady state recombination involves an equal number of holes and electrons. Therefore, the recombination rate of the majority carriers depends on the excess-minority-carrier-density as the minority carriers limit the recombination rate.

Recombination in a depletion region and in situations where the hole and electron density are close to each other cannot be described with the simple model and the more elaborate expressions for the individual recombination mechanisms must be used.

BAND-TO-BAND RECOMBINATION

Band-to-band recombination depends on the density of available electrons and holes. Both carrier types need to be available in the recombination process. Therefore, the rate is expected to be proportional to the product of n and p. Also, in thermal equilibrium, the recombination rate must equal the generation rate since there is no net recombination or generation. As the product of n and p equals n_i^2 in thermal equilibrium, the net recombination rate can be expressed as:

$$U_{b-b} = b\left(np - n_i^2\right)$$

where b is the bimolecular recombination constant.

TRAP ASSISTED RECOMBINATION

The net recombination rate for trap-assisted recombination is given by:

$$U_{SHR} = \frac{pn - n_i^2}{p + n + 2n_i \cosh(\frac{E_i - E_t}{kT})} N_t v_{th} \sigma$$

The derivation of this equation is beyond the scope of this text.

This expression can be further simplified for $p >> n$ to:

$$U_n = R_n - G_n = \frac{n_p - n_{p0}}{\tau_n}$$

and for $n >> p$ to:

$$U_p = R_p - G_p = \frac{p_n - p_{n0}}{\tau_p}$$

were

$$\tau_n = \tau_p = \frac{1}{N_t v_{th} \sigma}$$

SURFACE RECOMBINATION

Recombination at surfaces and interfaces can have a significant impact on the behaviour of semiconductor devices. This is because surfaces and interfaces typically contain a large number of recombination centres because of the abrupt termination of the semiconductor crystal, which leaves a large number of electrically active states. In addition, the surfaces and interfaces are more likely to contain impurities since they are exposed during the device fabrication process. The net recombination rate due to trap-assisted recombination and generation is given by:

$$U_s, SHR = \frac{pn - n_i^2}{p + n + 2n_i \cosh\left(\frac{E_i - E_{st}}{kT}\right)} N_{st} v_{th} \sigma_s$$

This expression is almost identical to that of Shockley-Hall-Read recombination. The only difference is that the recombination is due to a two-dimensional density of traps, N_{ts}, as the traps only exist at the surface or interface.

This equation can be further simplified for minority carriers in a quasi-neutral region. For instance for electrons in a quasi-neutral p-type region $p >> n$ and $p >> n_i$ so that for $E_i = E_{st}$, it can be simplified to:

$$U_{s,n} = R_{s,n} - G_{s,n} = v_s(n_p - n_{p0})$$

where the recombination velocity, v_s, is given by:

$$V_s = N_{st} v_{th} \sigma_s$$

AUGER RECOMBINATION

Auger recombination involves three particles: an electron and a hole, which recombine in a band-to-band transition and give off the resulting energy to another electron or hole. The expression for the net recombination rate is therefore similar to that of band-to-band recombination but includes the density of the electrons or holes, which receive the released energy from the electron-hole annihilation:

$$U_{Auger} = \Gamma_p p\left(np - n_i^2\right)$$

The two terms correspond to the two possible mechanisms.

GENERATION DUE TO LIGHT

Carriers can be generated in semiconductors by illuminating the semiconductor with light. The energy of the incoming photons is used to bring an electron from a lower energy level to a higher energy level. In the case where an electron is removed from the valence band and added to the conduction band, an electron-hole pair is generated. A necessary condition is that the energy of the photon, E_{ph}, is larger than the bandgap energy, E_g. As the energy of the photon is given off to the electron, the photon no longer exists.

If each absorbed photon creates one electron-hole pair, the electron and hole generation rates are given by:

$$G_{p,light} = G_{n,light} = \alpha \frac{P_{opt}(x)}{E_{ph}A}$$

where α is the absorption coefficient of the material at the energy of the incoming photon. The absorption of light in a semiconductor causes the optical power to decrease with distance. This effect is described mathematically by:

$$\frac{dP_{opt}(x)}{dx} = -\alpha P_{opt}(x)$$

The calculation of the generation rate of carriers therefore requires first a calculation of the optical power within the structure from which the generation rate can then be obtained using equation.

Example

Calculate the electron and hole densities in an n-type silicon wafer (N_d = 10^{17} cm^{-3}) illuminated uniformly with 10 mW/cm^2 of red light (E_{ph} = 1.8 eV). The absorption coefficient of red light in silicon is 10^{-3} cm^{-1}. The minority carrier lifetime is 10 μs.SolutionThe generation rate of electrons and holes equals:

$$G_n = G_p = \alpha \frac{P_{opt}}{E_{ph}A} = 10^{-3} \frac{10^{-2}}{1.8 \times 1.6 \times 10^{-19}} = 3.5 \times 10^{13}\ cm^{-3}s^{-1}$$

where the photon energy was converted into Joules. The excess carrier densities are then obtained from:

$$\delta n = \delta p = \zeta_p G_p = 10 \times 10^{-6} \times 3.5 \times 10^{8}\ cm^{-3}\ s^{-1}$$

The excess carrier densities are then obtained from: So that the electron and hole densities equal:

$$n = n_0 + \delta_n = 10^{17} + 3.5 \times 10^{13} = 10^{17}\ cm^{-3}\ s^{-1}$$

CONTINUITY EQUATION

DERIVATION

The continuity equation describes a basic concept, namely that a change in carrier density over time is due to the difference between the incoming and outgoing flux of carriers plus the generation and minus the recombination. The flow of carriers and recombination and generation rates are illustrated with Figure.

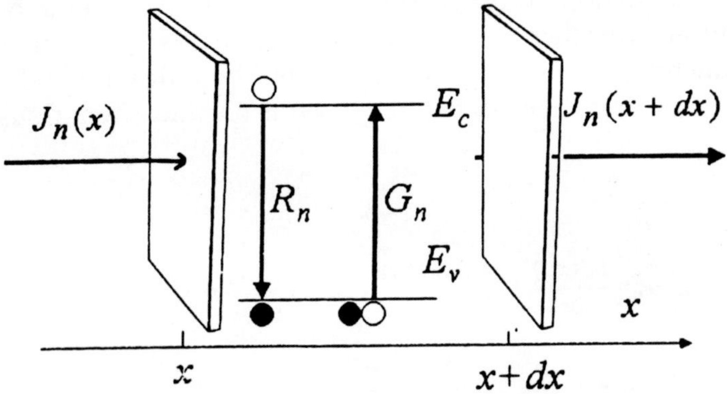

Fig. Electron currents and possible recombination and generation processes

The rate of change of the carriers between x and $x + dx$ equals the difference between the incoming flux and the outgoing flux plus the generation and minus the recombination:

$$\frac{\partial n(x,t)}{\partial t} A dx = \left(\frac{J_n(x)}{-q} - \frac{J_n(x+dx)}{-q} \right) A + \left(G_n(x,t) - R_n(x,t) \right) A dx$$

where $n(x,t)$ is the carrier density, A is the area, $G_n(x,t)$ is the generation rate and $R_n(x,t)$ is the recombination rate. Using a Taylor series expansion,

$$J_n(x+dx) = J_n(x) + \frac{d J_n(x)}{dx} dx$$

this equation can be formulated as a function of the derivative of the current:

$$\frac{\partial n(x,t)}{\partial t} = \frac{1}{q}\frac{\partial J_n(x,t)}{\partial x} + G_n(x,t) - R_{n(x,t)}$$

and similarly for holes one finds:

$$\frac{\partial n(x,t)}{\partial t} = -\frac{1}{q}\frac{\partial J_n(x,t)}{\partial x} + G_p(x,t) - R_{p(x,t)}$$

A solution to these equations can be obtained by substituting the expression for the electron and hole current.This then yields two partial differential equations as a function of the electron density, the hole density and the electric field. The electric field itself is obtained from Gauss's law.

$$\frac{\partial n(x,t)}{\partial t} =$$

$$\mu_n n \frac{\partial \mathcal{E}(x,t)}{\partial x} + \mu_n \mathcal{E} \frac{\partial n(x,t)}{\partial x} + D_n \frac{\partial^2 n(x,t)}{\partial x^2} + G_n(x,t) - R_n(x,t)$$

$$\frac{\partial p(x,t)}{\partial t} =$$

$$-\mu_p p \frac{\partial \mathcal{E}(x,t)}{\partial x} - \mu_p \mathcal{E} \frac{\partial p(x,t)}{\partial x} + D_p \frac{\partial^2 p(x,t)}{\partial x^2} + G_p(x,t) - R_p(x,t)$$

A generalization in three dimensions yields the following continuity equations for electrons and holes:

$$\frac{\partial n(x,y,z,t)}{\partial t} = \frac{1}{q}\vec{\nabla}\vec{J}_n(x,y,z,t) + G_n(x,y,z,t) - R_n(x,y,z,t)$$

$$\frac{\partial p(x,y,z,t)}{\partial t} = -\frac{1}{q}\vec{\nabla}\vec{J}_p(x,y,z,t) + G_p(x,y,z,t) - R_p(x,y,z,t)$$

THE DIFFUSION EQUATION

In the quasi-neutral region – a region containing mobile carriers, where the electric field is small - the current is due to diffusion only. In addition, we can use the simple recombination model for the net recombination rate since the recombination rates depend only on the minority carrier density. This leads to the time-dependent diffusion equations for electrons in p-type material and for holes in n-type material:

$$\frac{\partial n(x,t)}{\partial t} = D_n \frac{\partial^2 n_p(x,t)}{\partial x^2} - \frac{n_p(x,t) - n_{p0}}{\tau_n}$$

$$\frac{\partial p(x,t)}{\partial t} = D_p \frac{\partial^2 p_n(x,t)}{\partial x^2} - \frac{p_n(x,t) - p_{n0}}{\tau_p}$$

STEADY STATE SOLUTION TO THE DIFFUSION EQUATION

In steady state, the partial derivatives with respect to time are zero, yielding:

$$0 = D_n \frac{d^2 n_p(x)}{dx^2} - \frac{n_p(x) - n_{p0}}{\tau_n}$$

$$0 = D_p \frac{d^2 p_n(x)}{dx^2} - \frac{p_n(x) - p_{n0}}{\tau_p}$$

The general solution to these second order differential equations are:

$$n_p(x \leq -x_p) = n_{p0} + Ce^{-(x+x_p)/L_n} + De^{(x+x_p)/L_n}$$

$$p_n(x \geq x_n) = p_{n0} + Ae^{-(x-x_n)/L_p} + Be^{(x-x_n)/L_p}$$

where L_n and L_p are the diffusion lengths given by:

$$L_n = \sqrt{D_n \tau_n}$$

$$L_p = \sqrt{D_p \tau_p}$$

The diffusion constants, D_n and D_p, are obtained using the Einstein relations. The diffusion equations can also be written as a function of the *excess>/i> carrier densities, δn and δp, which are related to the total carrier densities, n and p, and the thermal equilibrium densities, n_0 and p_0, by:*

$$n = n_0 + \delta n$$

$$p = p_0 + \delta p$$

yielding:

$$0 = \frac{d^2(\delta n_p)}{dx^2} - \frac{\delta n_p}{L_n^2}$$

$$0 = \frac{d^2(\delta p_n)}{dx^2} - \frac{\delta p_n}{L_p^2}$$

The diffusion equation will be used to calculate the diffusion current in p-n junctions and bipolar transistors.

THE DRIFT-DIFFUSION MODEL

The drift-diffusion model of a semiconductor is frequently used to describe semiconductor devices. It contains all the features described in this chapter.

We will apply the drift-diffusion model to a variety of different devices. To facilitate this analysis, we present here a simplified drift-diffusion model, which contains all the essential features. This model results in a set of ten variables and ten equations.

The assumptions of the simplified drift-diffusion model are:

Full ionization: all dopants are assumed to be ionized (shallow dopants)

Non-degenerate: the Fermi energy is assumed to be at least 3 *kT* below/above the conduction/valence band edge.

Steady state: All variables are independent of time.

Constant temperature: The temperature is constant throughout the device.

The ten variables are the following:

ρ, the charge density

n, the electron density

p, the hole density

ε, the electric field

ϕ, the potential

E_i, the intrinsic energy

F_n, the electron quasi-Fermi energy

F_p, the hole quasi-Fermi energy

J_n, the electron current density

J_p, the hole current density

The ten equations are:

Charge density equation

$$\rho = q(p - n + N_d^+ - N_a^-)$$

Electric field and potential equations

$$\frac{d\varepsilon}{dx} = \frac{\rho}{\varepsilon}$$

$$\frac{d\phi}{dx} = -\varepsilon$$

$$\frac{dE_i}{dx} = q\varepsilon$$

Carrier density equations

$$n = n_i e^{(F_n - E_i)/kT}$$

$$p = n_i e^{(E_i - F_p)/kT}$$

Drift and diffusion current equations

$$J_n = qn\mu_n\varepsilon + qD_n\frac{dn}{dx}$$

$$J_p = qp\mu_p\varepsilon - qD_p\frac{dp}{dx}$$

Continuity equation in steady state with SHR recombination

$$0 = \frac{1}{q}\frac{\partial J_n}{\partial x} - \frac{np - n_i^2}{n + p + 2n_i \cosh(\frac{E_t - E_i}{kT})}\frac{1}{\tau}$$

$$0 = -\frac{1}{q}\frac{\partial J_p}{\partial x} - \frac{np - n_i^2}{n + p + 2n_i \cosh(\frac{E_t - E_i}{kT})}\frac{1}{\tau}$$

SEMICONDUCTOR THERMODYNAMICS

Thermodynamics can be used to explain some characteristics of semiconductors and semiconductor devices, which can not readily be explained based on the transport of single particles. One example is the fact that the Fermi energy is located within the energy gap where there are no energy levels and therefore also no electrons or holes. This is because the Fermi energy describes the energy of the particles in the distribution and the properties of the distribution can be markedly different from that of an individual particle. Thermodynamics can also be used to gain a completely different perspective. This section is therefore a worthwhile one to study as one can gain further insight into the field of semiconductors. However, it is not required to understand the following chapters.

THERMAL EQUILIBRIUM

A system is in thermal equilibrium if detailed balance is obtained. Detailed balance implies that every process in the system is exactly balanced by its inverse process. As a result, there is no net effect on the system.

This definition implies that in thermal equilibrium no energy (heat, work or particle energy) is exchanged between the parts within the system and between the system and the environment. Thermal equilibrium is obtained by isolating a system from its environment, removing any internal sources of energy, and waiting for a long enough time until the system does not change any more.

The concept of thermal equilibrium is of interest since a variety of thermodynamic results assume that the system under consideration is in thermal equilibrium. Few systems of interest rigorously satisfy this condition so that we often apply the thermodynamical results to systems, which are "close" to thermal equilibrium. Agreement between theories based on this assumption and experiments justifies this approach.

THERMODYNAMIC IDENTITY

The thermodynamic identity simply states that adding heat, work or particles can cause a change in energy. Mathematically this is expressed by:

$$dU = dQ + dW + \mu dN$$

where U is the energy, Q is the heat and W is the work. μ is the energy added to a system when adding one particle without adding either heat or work. The amount of exchanged heat depends on the temperature, T, and the entropy, S, while the amount of work delivered to a system depends on the pressure, p, and the volume, V, or:

$$dQ = TdS$$

and

$$dW = -pdV$$

yielding:

$$dU = TdS - pdV + \mu dN$$

THE FERMI ENERGY

The Fermi energy, E_F, is the energy associated with a particle, which is in thermal equilibrium with the system of interest. The energy is strictly associated with the particle and does not consist even in part of heat or work. This same quantity is called the electro-chemical potential, μ, in most thermodynamics texts.

EXAMPLE: AN IDEAL ELECTRON GAS

To illustrate the difference between the average energy of particles in a system and the Fermi energy, we now consider an ideal electron gas. The term ideal refers to the fact that the gas obeys the ideal gas law. To be "ideal" the gas must consist of particles, which do not interact with each other.

The total energy of the non-degenerate electron gas containing N particles equals:

$$U = \frac{3}{2}NkT + NE_c$$

as each non-relativistic electron has a thermal energy of $kT/2$ for each degree of freedom in addition to its minimum energy, E_c. The product of the pressure and volume is given by the ideal gas law:

$$pV = NkT$$

While the Fermi energy is given by equation:

$$E_F = E_c + kT \ln \frac{n}{N_c}$$

The thermodynamic identity can now be used to find the entropy from:

$$S = \frac{U - \mu N + pV}{T}$$

yielding:

$$S = \frac{5}{2}Nk - Nk \ln \frac{n}{N_c}$$

This relation can be visualized on an energy band diagram when one considers the energy, work and entropy per electron and compares it to the electro-chemical potential as shown in Figure.

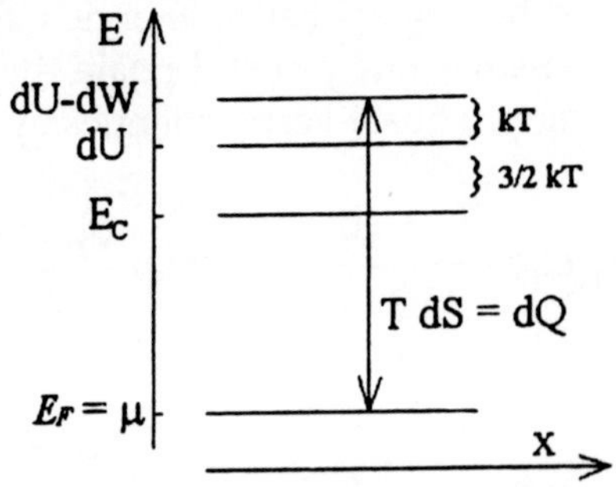

Fig. Energy, work and heat per electron in an ideal electron gas visualized on an energy band diagram.

The distinction between the energy and the electrochemical potential also leads to the following observations: Adding more electrons to an ideal electron gas with an energy which equals the average energy of the electrons in the gas increases both the particle energy and the entropy as heat is added in addition to particles. On the other hand, when bringing in electrons through an electrical contact whose voltage equals the Fermi energy (in electron volts) one does not add heat and the energy increase equals the Fermi energy times the number of electrons added.

Therefore, when analyzing the behaviour of electrons and holes on an energy band diagram, one should be aware of the fact that the total energy of an electron is given by its position on the diagram, but that the particle energy is given by the Fermi energy. The difference is the heat minus the work per electron or $dQ - dW = T\,dS + p\,dV$.

QUASI-FERMI ENERGIES

Quasi-Fermi energies are introduced when the electrons and holes are clearly not in thermal equilibrium with each other. This occurs when an external voltage is applied to the device of interest. The quasi-Fermi energies are introduced based on the notion that even though the electrons and holes are not in thermal equilibrium with each other, they still are in thermal equilibrium with themselves and can still be described by a Fermi energy which is now different for the electrons and the holes. These Fermi energies are referred to as the electron and hole quasi-Fermi energies, F_n and F_p. For non-degenerate densities one can still relate the electron and hole densities to the two quasi-Fermi energies by the following equations:

$$n = n_i \exp\frac{F_n - E_i}{kT} = N_c \exp\frac{F_n - E_c}{kT}$$

$$p = n_i \exp\frac{E_i - F_p}{kT} = N_v \exp\frac{E_v - F_n}{kT}$$

ENERGY LOSS IN RECOMBINATION PROCESSES

The energy loss in a recombination process equals the difference between the electron and hole quasi-Fermi energies as the energy loss is only due to the energy of the particles, which are lost:

$$\Delta U = F_n - F_p$$

No heat or work is removed from the system, just the energy associated with the particles. The energy lost in the recombination process can be converted in heat or light depending on the details of the process.

THERMO-ELECTRIC EFFECTS IN SEMICONDUCTORS

The temperature dependence of the current in a semiconductor can be included by generalizing the drift-diffusion current equation. The proportionality constant between the current density and the temperature gradient is the product of the conductivity, σ, and the thermo-electric power, π.

The derivation starts by generalizing the diffusion current to include a possible variation of the diffusion constant with position, yielding:

$$J_n = q\mu_n n\varepsilon + q\frac{d(D_n n)}{dx}$$

If the semiconductor is non-degenerate the electron density can be related to the effective density of states and the difference between the Fermi energy and the conduction band edge:

$$n = N_c \exp \frac{E_F - E_c}{kT}$$

yielding:

$$J_n = \mu_n n \left(\frac{dE_c}{dx} + k\frac{dT}{dx} + \frac{kT}{\mu_n}\frac{d\mu_n}{dx} + \frac{kT}{N_c}\frac{dN_c}{dx} + kT\frac{d\left(\frac{E_F - E_c}{kT}\right)}{dx} \right)$$

For the case where the material properties do not change with position, all the spatial variations except for the gradient of the Fermi energy are caused by a temperature variation. We postulate that the current can be written in the following form:

$$J_n = \mu_n n \left(\frac{dE_c}{dx} - qp\frac{dT}{dx} \right)$$

and π is the thermo-electric power in Volt/Kelvin. From both equations one then obtains the thermo-electric power:

$$P_n = -\frac{k}{q}\left(\frac{5}{2} + \frac{T}{\mu_n}\frac{d\mu_n}{dT} + \text{In}\frac{N_c}{n} \right)$$

If the temperature dependence of the mobility can be expressed as a simple power law:

$$\mu_n \alpha T^{-s}$$

the thermo-electric power becomes:

$$P_n = -\frac{k}{q}\left(\frac{5}{2} - s + \text{In}\frac{N_c}{n} \right)$$

for n-type material and similarly for p-type material:

$$P_p = -\frac{k}{q}\left(\frac{5}{2} - s + \text{In}\frac{N_v}{p} \right)$$

The Peltier coefficient, π, is related to the thermo-electric power by:

$$\Pi = pT$$

If electrons and holes are present in the semiconductor one has to include the effect of both when calculating the Peltier coefficient:

$$\Pi_{\text{total}} = \frac{\sigma_n \Pi_n + \sigma_p \Pi_p}{\sigma_n + \sigma_p}$$

The resulting Peltier coefficient as a function of temperature for silicon is shown in the figure below:

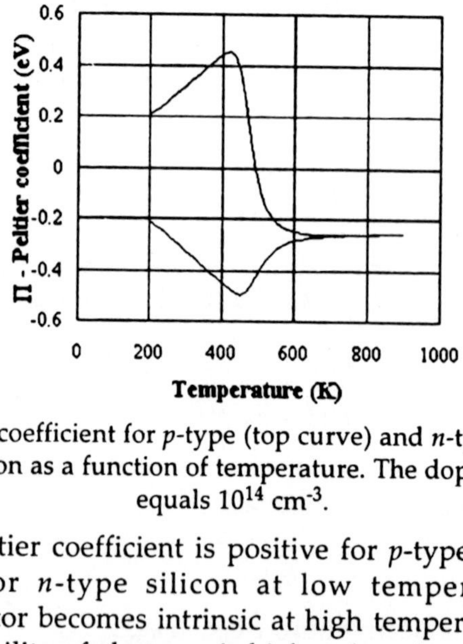

Fig. Peltier coefficient for *p*-type (top curve) and *n*-type (bottom curve) silicon as a function of temperature. The doping density equals 10^{14} cm^{-3}.

The Peltier coefficient is positive for *p*-type silicon and negative for *n*-type silicon at low temperature. The semiconductor becomes intrinsic at high temperature. Given that the mobility of electrons is higher than that of holes, the Peltier coefficient of intrinsic silicon is negative.

THE THERMO-ELECTRIC COOLER

Thermo-electric effects in semiconductors cause carriers to flow due to temperature gradients but also cause temperature gradients when an electrical current is applied. The thermo-electric cooler is a practical device in which a current is applied to a semiconductor causing a temperature reduction and cooling.

Such thermo-electric cooler consists of multiple semiconductor elements, which are connected in series as shown in Figure. The doping density in the semiconductor elements is graded with the highest density at the high temperature end and the low density at the low temperature end. An electrical current is applied to the series connection of these elements. Alternating *n*-type and *p*-type elements are used to ensure that the carriers flow in the same direction. While in principle a single piece of semiconducting material could have been used, the series connection is typically chosen to avoid the high current requirement of the single element.

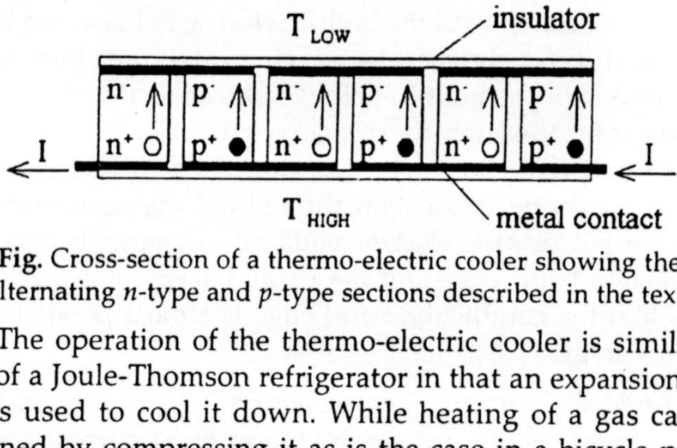

Fig. Cross-section of a thermo-electric cooler showing the alternating *n*-type and *p*-type sections described in the text.

The operation of the thermo-electric cooler is similar to that of a Joule-Thomson refrigerator in that an expansion of a gas is used to cool it down. While heating of a gas can be obtained by compressing it as is the case in a bicycle pump (where some of the heating is due to friction), a gas can also be cooled by expanding it into a larger volume. This process is most efficient if no heat is exchanged with the environment. This is also referred to as an isentropic expansion since the entropy remains constant if no heat is exchanged.

The gas in a thermo-electric cooler is the electron or hole gas. As a current is applied such that carriers flow from the high density (high *T*) region to the low-density (low *T*) region, one can imagine that the volume around a fixed number of carriers must increase as the carriers move towards the lower doped region. A possible energy band diagram is shown below:

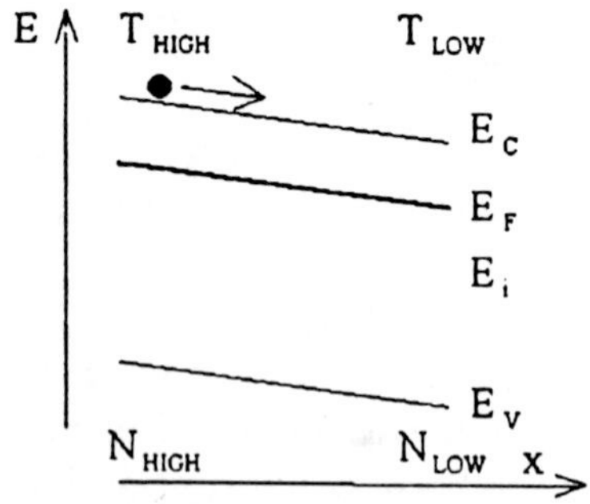

Fig. Energy band diagram of an n-type element of the thermo electric cooler of Figure

At constant temperature and in thermal equilibrium there is no current as the diffusion current is balanced by the drift current associated with the built-in electric field caused by the graded doping density. As a current is applied to the semiconductor the built-in field is reduced so that the carriers diffuse from the high to low doping density. This causes a temperature reduction on the low-doped side, which continues until the entropy is constant throughout the semiconductor. Since the entropy per electron equals the distance between the conduction band edge and the Fermi energy plus 5/2 kT one finds that the conduction band edge is almost parallel to the Fermi energy.

An ideal isentropic expansion is typically not obtained due to the Joule heating caused by the applied current and the thermal losses due to the thermal conductivity of the material. The need to remove heat at the low temperature further increases the lowest achievable temperature.

THE "HOT-PROBE" EXPERIMENT

The "hot-probe" experiment provides a very simple way to distinguish between n-type and p-type semiconductors using a soldering iron and a standard multi-meter.

The experiment is performed by contacting a semiconductor wafer with a "hot" probe such as a heated soldering iron and a "cold" probe. Both probes are wired to a sensitive current meter. The hot probe is connected to the positive terminal of the meter while the cold probe is connected to the negative terminal. The experimental set-up is shown in Figure:

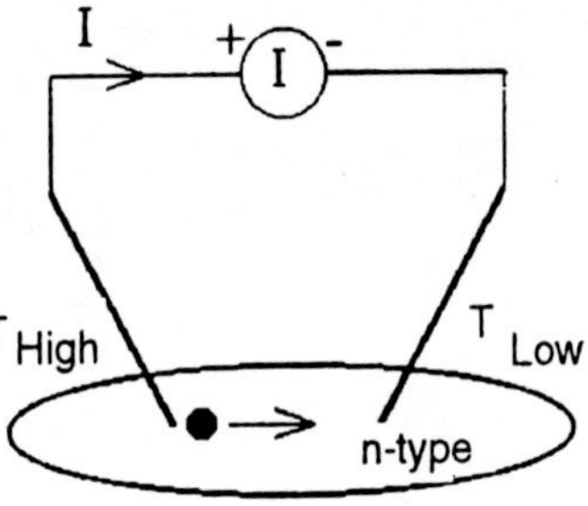

Fig. Experimental set-up of the "hot-probe" experiment.

When applying the probes to *n*-type material one obtains a positive current reading on the meter, while *p*-type material yields a negative current.

A simple explanation for this experiment is that the carriers move within the semiconductor from the hot probe to the cold probe. While diffusion seems to be a plausible mechanism to cause the carrier flow it is actually not the most important mechanism since the material is uniformly doped. However, as will be discussed below there is a substantial electric field in the semiconductor so that the drift current dominates the total current.

Starting from the assumption that the current meter has zero resistance, and ignoring the (small) thermoelectric effect in the metal wires one can justify that the Fermi energy does not vary throughout the material. A possible corresponding energy band diagram is shown below:

This energy band diagram illustrates the specific case in which the temperature variation causes a linear change of the conduction band energy as measured relative to the Fermi energy, and also illustrates the trend in the general case. As the effective density of states decreases with decreasing temperature, one finds that the conduction band energy decreases with decreasing temperature yielding an electric field, which causes the electrons to flow from the high to the low temperature. The same reasoning reveals that holes in a p-type semiconductor will also flow from the higher to the lower temperature.

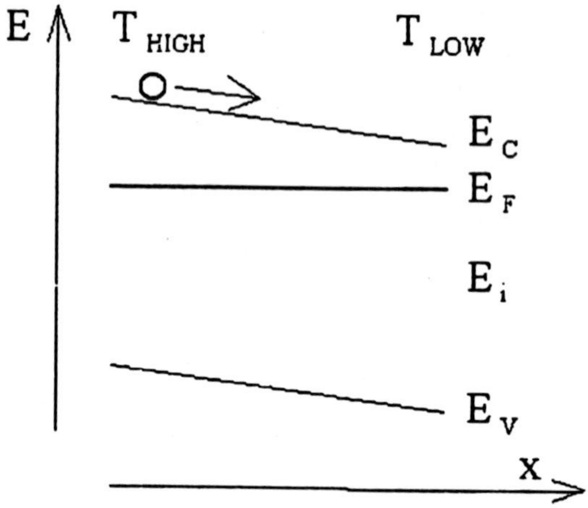

Fig. Energy band diagram corresponding to the "hot-probe" experiment illustrated by Figure.

The current can be calculated from the general expression:

$$J_n = \mu_n n \left(\frac{dE_c}{dx} - qP \frac{dT}{dx} \right)$$

where

$$qP_n = -k \left(\frac{5}{2} + \frac{T}{\mu_n} \frac{d\mu_n}{dx} + \text{In} \frac{N_c}{n} \right)$$

The current will therefore increase with doping and with the applied temperature gradient as long as the semiconductor does not become degenerate or intrinsic within the applied temperature range.

Chapter 2

Metal-Semicond Junctions

INTRODUCTION

Metal-to-semiconductor (M-S) junctions are of great importance since they are present in every semiconductor device. They can behave either as a Schottky barrier or as an ohmic contact dependent on the characteristics of the interface. We will primarily focus on Schottky barriers. This chapter contains an analysis of the electrostatics of the M-S junction, including the calculated of the charge, field and potential distribution within the device. This chapter also contains a derivation of the current voltage characteristics due to diffusion, thermionic emission and tunneling in Metal-Semiconductor junctions, followed by a discussion of M-S contacts and MESFETs.

STRUCTURE AND PRINCIPLE OF OPERATION

STRUCTURE

The structure of a metal-semiconductor junction is shown in Figure. It consists of a metal contacting a piece of semiconductor. An ideal Ohmic contact, a contact such that no potential exists between the metal and the semiconductor, is made to the other side of the semiconductor. The sign convention of the applied voltage and current is also shown on Figure.

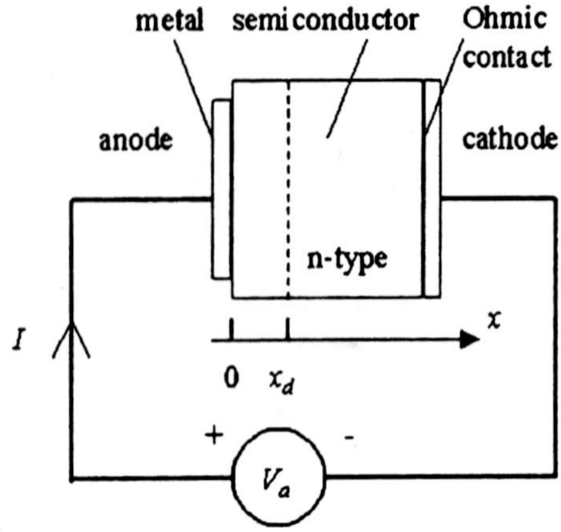

Fig. Structure and sign convention of a metal-semiconductor junction

FLATBAND DIAGRAM AND BUILT-IN POTENTIAL

The barrier between the metal and the semiconductor can be identified on an energy band diagram. To construct such diagram we first consider the energy band diagram of the metal and the semiconductor, and align them using the same vacuum level as shown in Figure. As the metal and semiconductor are brought together, the Fermi energies of the metal and the semiconductor do not change right away. This yields the flatband diagram of Figure.

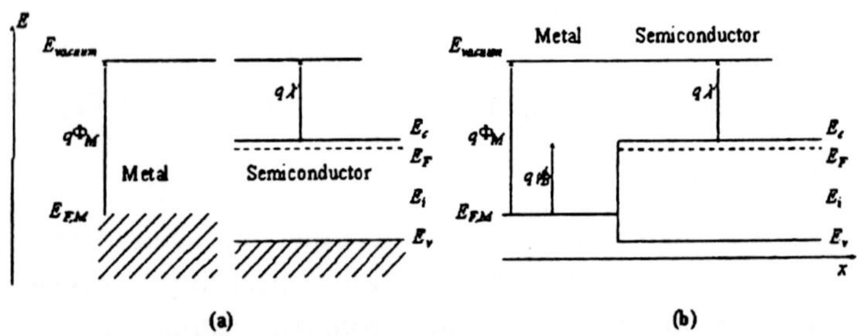

Fig. Energy band diagram of the metal and the semiconductor before (a) and after (b) contact is made.

The barrier height, ϕ_B, is defined as the potential difference between the Fermi energy of the metal and the band edge where the majority carriers reside. From Figure one finds that for an *n*-type semiconductor the barrier height is obtained from:

$$\phi_B = \phi_M - \chi, \text{ for an n-type semiconductor}$$

Where Φ_M is the work function of the metal and χ is the electron affinity. The work function of selected metals as measured in vacuum can be found in Table. For *p*-type material, the barrier height is given by the difference between the valence band edge and the Fermi energy in the metal:

$$\phi_B = \frac{E_g}{q} + \chi - \phi_M, \text{ for an p-type semiconductor}$$

A metal-semiconductor junction will therefore form a barrier for electrons and holes if the Fermi energy of the metal as drawn on the flatband diagram is somewhere between the conduction and valence band edge.

In addition, we define the built-in potential, ϕ_I, as the difference between the Fermi energy of the metal and that of the semiconductor.

$$\phi_I = \phi_M - \chi - \frac{E_c - E_{F,}n}{q}, \text{ n-type}$$

$$\phi_I = \chi + \frac{E_c - E_{F,}p}{q} - \phi_M, \text{ p-type}$$

The measured barrier height for selected metal-semiconductor junctions is listed in Table. These experimental barrier heights often differ from the ones calculated using equation. This is due to the detailed behaviour of the metal-semiconductor interface. The ideal metal-semiconductor theory assumes that both materials are infinitely pure, that there is no interaction between the two materials and no unwanted interfacial layer. Chemical reactions between the metal and the semiconductor alter the barrier height as do interface states at the surface of the semiconductor and interfacial layers. Some general trends however can still be observed.The barrier height on *n*-type semiconductors increases for metals with a higher work function as can be verified for silicon. Gallium arsenide on the other hand is known to have a large density of surface states so that the barrier height becomes virtually independent of the metal. Furthermore, one finds that the barrier heights reported in the literature to vary widely due to different surface cleaning procedures.

	Ag	Al	Au	Cr	Ni	Pt	W
ϕM (in vacuum)	4.3	4.25	4.8	4.5	4.5	5.3	4.6
n-Ge	0.54	0.48	0.59		0.49		0.48
p-Ge	0.5		0.3				
n-Si	0.78	0.72	0.8	0.61	0.61	0.9	0.67
p-Si	0.54	0.58	0.34	0.5	0.51		0.45
n-GaAs	0.88	0.8	0.9			0.84	0.8
p-GaAs	0.63		0.42				

Workfunction in units of eV of selected metals and their measured barrier height on germanium, silicon and gallium arsenide.

Example

Consider a chrome-silicon metal-semiconductor junction with $N_d = 10^{17}$ cm^{-3}. Calculate the barrier height and the built-in potential. Repeat for a *p*-type semiconductor with the same doping density.

Solution

The barrier height equals:

$$\phi_B = \Phi_M - \chi = 4.5 - 4.05 = 0.45 \text{ V}$$

Note that this value differs from the one listed in Table, since the work function in vacuum was used. See the discussion in the text for more details.

The built-in potential equals:

$$\phi_i = \phi_B - V_t \ln \frac{N_c}{N_d} = 0.45 - 0.0259 \ln \frac{2.82 \times 10^{19}}{10^{17}} = 0.30 \text{ V}$$

The barrier height for the chrome/*p*-silicon junction equals:

$$\phi_B = \chi + \frac{E_g}{q} - \Phi_M = 4.05 + 1.12 - 4.5 = 0.67\text{ V}$$

And the built-in potential equals:

$$\phi_i = \phi_B - V_t \ln\frac{N_v}{N_a} = 0.67 - 0.0259\ln\frac{1.83\times10^{19}}{10^{17}} = 0.53\text{ V}$$

THERMAL EQUILIBRIUM

The flatband diagram, shown in Figure, is not a thermal equilibrium diagram, since the Fermi energy in the metal differs from that in the semiconductor. Electrons in the *n*-type semiconductor can lower their energy by traversing the junction. As the electrons leave the semiconductor, a positive charge, due to the ionized donor atoms, stays behind. This charge creates a negative field and lowers the band edges of the semiconductor. Electrons flow into the metal until equilibrium is reached between the diffusion of electrons from the semiconductor into the metal and the drift of electrons caused by the field created by the ionized impurity atoms. This equilibrium is characterized by a constant Fermi energy throughout the structure.

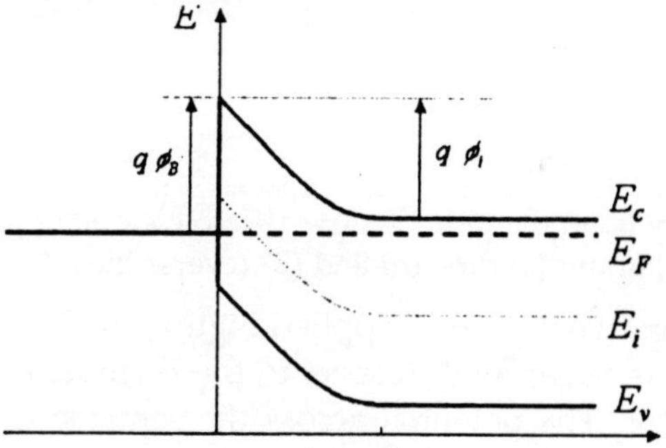

Fig. Energy band diagram of a metal-semiconductor contact in thermal equilibrium.

It is of interest to note that in thermal equilibrium, i.e. with no external voltage applied, there is a region in the semiconductor close to the junction ( ),which is depleted of mobile carriers. We call this the depletion region. The potential across the semiconductor equals the built-in potential, ϕ_i.

FORWARD AND REVERSE BIAS

Operation of a metal-semiconductor junction under forward and reverse bias is illustrated with Figure. As a positive bias is applied to the metal, the Fermi energy of the metal is lowered with respect to the Fermi energy in the semiconductor. This results in a smaller potential drop across the semiconductor. The balance between diffusion and drift is disturbed and more electrons will diffuse towards the metal than the number drifting into the semiconductor. This leads to a positive current through the junction at a voltage comparable to the built-in potential.

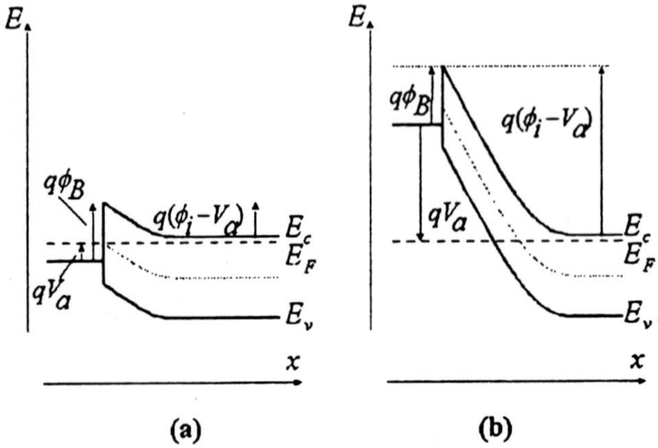

Fig. Energy band diagram of a metal-semiconductor junction under (a) forward and (b) reverse bias

As a negative voltage is applied Figure, the Fermi energy of the metal is raised with respect to the Fermi energy in the semiconductor. The potential across the semiconductor now increases, yielding a larger depletion region and a larger electric field at the interface. The barrier, which restricts the electrons to the metal, is unchanged so that that barrier, independent of the applied voltage, limits the flow of electrons. The metal-semiconductor junction with positive barrier height has therefore a pronounced rectifying behaviour. A large current exists under forward bias, while almost no current exists under reverse bias.

The potential across the semiconductor therefore equals the built-in potential, ϕ_i, minus the applied voltage, V_a.

$$\phi(x=\infty)-\phi(x=0)=\phi_1-V_a$$

ELECTROSTATIC ANALYSIS

GENERAL DISCUSSION - POISSON'S EQUATION

The elcctrostatic analysis of a metal-semiconductor junction is of interest since it provides knowledge about the charge and field in the depletion region. It is also required to obtain the capacitance-voltage characteristics of the diode.

The general analysis starts by setting up Poisson's equation:

$$\frac{d^2\phi}{dx^2}=-\frac{\rho}{E_s}=-\frac{q}{E_s}\left(p-n+N_a^+-N_a^-\right)$$

Where the charge density, ρ, is written as a function of the electron density, the hole density and the donor and acceptor densities. To solve the equation, we have to express the electron and hole density, n and p, as a function of the potential, ϕ, yielding:

$$\frac{d^2\phi}{dx^2}=\frac{2qn_i}{E_s}\left(\sinh\frac{\phi-\phi_F}{V_t}+\sinh\frac{\phi_F}{V_t}\right)$$

with

$$\sinh\frac{\phi_F}{V_t}=\frac{N_a^- - N_a^+}{2n_i}$$

where the potential is chosen to be zero in the n-type region, where $x \gg x_n$.

This second-order non-linear differential equation cannot be solved analytically. Instead we will make the simplifying assumption that the depletion region is fully depleted and that the adjacent neutral regions contain no charge.

FULL DEPLETION APPROXIMATION

The simple analytic model of the metal-semiconductor junction is based on the full depletion approximation. This approximation is obtained by assuming that the semiconductor is fully depleted over a distance x_d, called the depletion region. While this assumption does not provide an accurate charge distribution, it does provide very reasonable approximate expressions for the electric field and potential throughout the semiconductor.

FULL DEPLETION ANALYSIS

We now apply the full depletion approximation to an M-S junction containing an *n*-type semiconductor. We define the depletion region to be between the metal-semiconductor interface ($x = 0$) and the edge of the depletion region ($x = x_d$). The depletion layer width, x_d, is unknown at this point but will later be expressed as a function of the applied voltage.

To find the depletion layer width, we start with the charge density in the semiconductor and calculate the electric field and the potential across the semiconductor as a function of the depletion layer width. We then solve for the depletion layer width by requiring the potential across the semiconductor to equal the difference between the built-in potential and the applied voltage, $\phi_i - V_a$. The different steps of the analysis are illustrated by Figure.

As the semiconductor is depleted of mobile carriers within the depletion region, the charge density in that region is due to the ionized donors. Outside the depletion region, the semiconductor is assumed neutral. This yields the following expressions for the charge density, ρ:

$$\rho(x) = qN_d \qquad 0 < x < x_d$$

$$\rho(x) = 0 \qquad x_d < x$$

where we assumed full ionization so that the ionized donor density equals the donor density, N_d. This charge density is shown in Figure. The charge in the semiconductor is exactly balanced by the charge in the metal, Q_M, so that no electric field exists except around the metal-semiconductor interface.

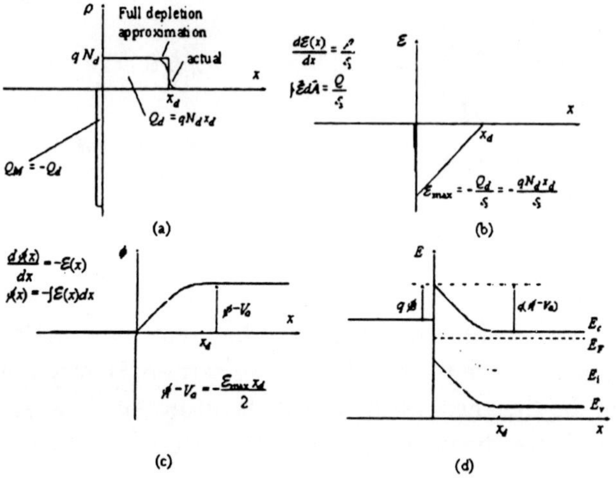

Fig. (a) Charge density, (b) electric field, (c) potential and (d) energy as obtained with the full depletion analysis.

Using Gauss's law we obtain the electric field as a function of position, also shown in Figure:

$$\mathcal{E}(x) = -\frac{qN_d}{\varepsilon_s}(x_d - x) \qquad 0 < x < x_d$$

$$\mathcal{E}(x) = 0 \qquad\qquad x_d \leq x$$

where ε_s is the dielectric constant of the semiconductor. We also assumed that the electric field is zero outside the depletion region, since a non-zero field would cause the mobile carriers to redistribute until there is no field. The depletion region does not contain mobile carriers so that there can be an electric field. The largest (absolute) value of the electric field is obtained at the interface and is given by:

$$\varepsilon(x=0) = -\frac{qN_d x_d}{\varepsilon_s} = -\frac{Q_d}{\varepsilon_s}$$

where the electric field was also related to the total charge (per unit area), Q_d, in the depletion layer. Since the electric field is minus the gradient of the potential, one obtains the potential by integrating the expression for the electric field, yielding:

$$\phi(x) = 0 \qquad x \leq 0$$

$$\phi(x) = \frac{qN_d}{2\varepsilon_s}[x_d^2 - (x_d - x)^2] \qquad 0 < x < x_d$$

$$\phi(x) = \frac{qN_d x_d^2}{2\varepsilon_s} \qquad x_d \leq x$$

We now assume that the potential across the metal can be neglected. Since the density of free carriers is very high in a metal, the thickness of the charge layer in the metal is very thin. Therefore, the potential across the metal is several orders of magnitude smaller than that across the semiconductor, even though the total amount of charge is the same in both regions.

The total potential difference across the semiconductor equals the built-in potential, ϕ_i,in thermal equilibrium and is further reduced/increased by the applied voltage when a positive/negative voltage is applied to the metal as described by equation. This boundary condition provides the following relation between the semiconductor potential at the surface, the applied voltage and the depletion layer width:

$$\phi_i - V_a = -\phi(x = 0) = \frac{qN_d x_d^2}{2\varepsilon_s}$$

Solving this expression for the depletion layer width, x_d, yields:

$$x_d = \sqrt{\frac{2\varepsilon_s(\phi_i - V_a)}{qN_d}}$$

JUNCTION CAPACITANCE

In addition, we can obtain the capacitance as a function of the applied voltage by taking the derivative of the charge with respect to the applied voltage yielding:

$$C_j = \left|\frac{dQ_d}{dV_a}\right| = \sqrt{\frac{qE_s N_d}{2(\phi_i - V_a)}} = \frac{E_s}{x_d}$$

The last term in the equation indicates that the expression of a parallel plate capacitor still applies. This can be explained by the fact that the charge added/removed from the depletion layer as one decreases/increases the applied voltage is added/ removed only at the edge of the depletion region. While the parallel plate capacitor expression seems to imply that the capacitance is constant, the metal-semiconductor junction capacitance is not constant since the depletion layer width, x_d, varies with the applied voltage.

Example

Consider a chrome-silicon metal-semiconductor junction with $N_d = 10^{17}$ cm^{-3}. Calculate the depletion layer width, the electric field in the silicon at the metal-semiconductor interface, the potential across the semiconductor and the capacitance per unit area for an applied voltage of -5 V.

Solution

The depletion layer width equals:

$$x_d = \sqrt{\frac{2\varepsilon_s(\phi_i - V_a)}{qN_d}}$$

$$= \sqrt{\frac{2\times 11.9\times 8.85\times 10^{-14}\times(0.3+5)}{1.6\times 10^{-19}\times 10^{17}}} = 0.26\ \mu\text{m}$$

where the built-in potential was already calculated in previous Example.

The electric field in the semiconductor at the interface is:

$$\mathcal{E}(x=0) = \frac{qN_d x_d}{\varepsilon_s}$$

$$= \frac{1.6\times10^{-19}\times10^{17}\times2.6\times10^{-5}}{11.9\times8.85\times10^{-14}} = 4.0\times10^{5}\ \text{V/cm}$$

The potential equals:

$$\phi(x = x_d) = \frac{qN_d x_d^2}{2\varepsilon_s} = \phi_i - V_a = 5.3V$$

And the capacitance per unit area is obtained from:

$$C_j = \frac{\varepsilon_s}{x_d} = \frac{11.9\times8.85\times10^{-14}}{2.6\times10^{-5}} = 40\ \text{nF/cm}^2$$

SCHOTTKY BARRIER LOWERING

Image charges build up in the metal electrode of a metal-semiconductor junction as carriers approach the metal-semiconductor interface. The potential associated with these charges reduces the effective barrier height. This barrier reduction tends to be rather small compared to the barrier height itself. Nevertheless this barrier reduction is of interest since it depends on the applied voltage and leads to a voltage dependence of the reverse bias current. Note that this barrier lowering is only experienced by a carrier while approaching the interface and will therefore not be noticeable in a capacitance-voltage measurement.

An energy band diagram of an *n*-type silicon Schottky barrier including the barrier lowering is shown in Figure:

Shown is the energy band diagram obtained using the full-depletion approximation, the potential reduction experienced by electrons, which approach the interface and the resulting conduction band edge. A rounding of the conduction band edge can be observed at the metal-semiconductor interface as well as a reduction of the height of the barrier.

The calculation of the barrier reduction assumes that the charge of an electron close to the metal-semiconductor interface attracts an opposite surface charge, which exactly balances the electron's charge so that the electric field surrounding the electron does not penetrate beyond this surface charge.

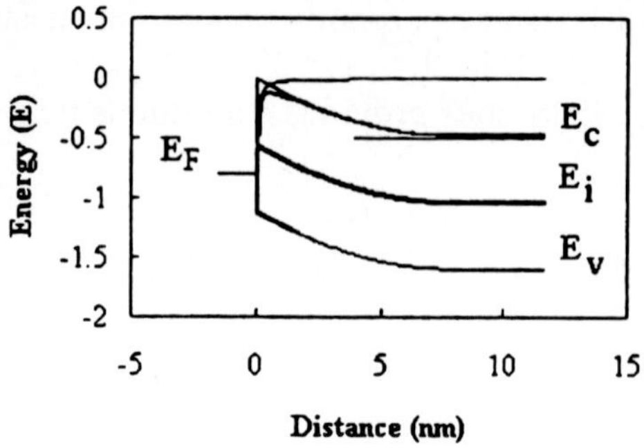

Fig. Energy band diagram of a silicon Schottky barrier with ϕ_B = 0.8 V and $N_d = 10^{19}$ cm^{-3}.

The time to build-up the surface charge and the time to polarize the semiconductor around the moving electron are assumed to be much shorter than the transit time of the electron . This scenario is based on the assumption that there are no mobile or fixed charges around the electron as it approaches the metal-semiconductor interface. The electron and the induced surface charges are shown in Figure:

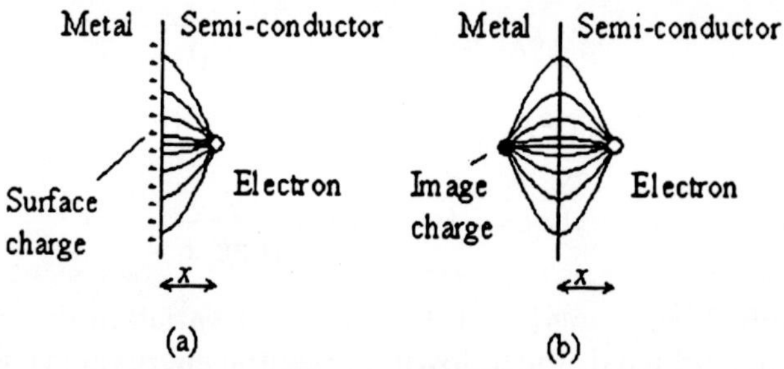

Fig. a) Field lines and surface charges due to an electron in close proximity to a perfect conductor and b) the field lines and image charge of an electron.

It can be shown that the electric field in the semiconductor is identical to that of the carrier itself and another carrier with opposite charge at equal distance but on the opposite side of the interface. This charge is called the image charge. The difference between the actual surface charges and the image charge is that the fields in the metal are distinctly different. The image charge concept is justified on the basis that the electric field lines are perpendicular to the surface a perfect conductor, so that, in the case of a flat interface, the mirror image of the field lines provides continuous field lines across the interface.

The barrier lowering depends on the square root of the electric field at the interface and is calculated from:

$$\Delta\phi_B = \sqrt{\frac{q\varepsilon_{\max}}{4\pi\varepsilon_s}}$$

DERIVATION OF SCHOTTKY BARRIER LOWERING

We now derive equation by calculating the potential due to the image charge and adding it to the potential within the depletion region.

The electrostatic force between the two particles, one with a positive electronic charge and the other with a negative electronic charge, which are both a distance, x, away from the interface ($x = 0$), is given by:

$$F(x) = q\varepsilon_I(x) = \frac{-q^2}{4\pi\varepsilon_s(2x)^2}$$

The corresponding potential equals:

$$\phi(x) = -\int_x^{\infty} \varepsilon_I(x)\,dx = \frac{q}{16\pi\varepsilon_s x}$$

which combined with the potential variation due to the electric field yields the following potential energy, $V(x)$, versus position, x:

$$V(x) = -q\varepsilon_{\max}x - \frac{q^2}{16\pi\varepsilon_s x}$$

where the field due to the charge in the depletion region is assumed to be constant and set equal to the maximum field, E_{max}.

$$\varepsilon_{\max} = \sqrt{\frac{qNd^2(\phi_i - V_a)}{\varepsilon_s}}$$

At this point the question arises why a single electron can noticeably alter the potential, while the depletion layer contains significantly more charge. To understand this, one has to realise that we have assumed that the charge in the depletion region is not quantized, but instead is distributed throughout the depletion layer. While this assumption does provide the correct average potential it does not accurately reflect the potential variations due to the individual charges of the ionized donors or mobile electrons. As we assumed that the single electron is far away from all other charges in the semiconductor, the potential energy due to all those charges is close to the average potential energy.

The potential energy due to the distributed charge of the ionized donors and a single electron reaches its maximum value at:

$$\varepsilon_{\max} = \sqrt{\frac{q}{16_{\pi\varepsilon_s}\varepsilon_{\max}}}$$

and the corresponding maximum value of the potential energy equals:

$$V_{\max} = q\phi_B - q\Delta\phi_B$$

where Df_B is the barrier height reduction given by:

$$\Delta\phi_B = \sqrt{\frac{q\varepsilon_{\max}}{4\pi\varepsilon_s}}$$

SOLUTION TO POISSON'S EQUATION

To assess the error made when using the full depletion approximation we now derive the correct solution by solving Poisson's equation analytically. The actual solution for the potential is then obtained by numerically integrating the expression for the electric field. We start from the charge density, ρ, in a semiconductor for the general case where electrons, holes, ionized acceptors and ionized donors are present:

$$\rho(\phi) = q\left(p + N_d^+ - n - N_a^-\right)$$

where ϕ is the potential in the semiconductor. The potential is chosen to equal zero deep into the semiconductor. For an *n*-type semiconductor without acceptors or free holes this can be further reduced to:

$$\rho(\phi) = qNd\left(1 - \exp\left(\frac{q\phi}{kT}\right)\right)$$

assuming the semiconductor to be non-degenerate and fully ionized. A similar expression can be obtained for *p*-type material. Poisson's law can then be rewritten as:

$$\frac{d^2\phi}{dx^2} = -\frac{\rho(\phi)}{\varepsilon_s} = -qNd\left(1 - \exp\left(\frac{q\phi}{kT}\right)\right)$$

Multiplying both sides with $d\phi/dx$, this equation can be integrated between an arbitrary point x and infinity. The electric field at infinity (deep in the semiconductor) is taken to be zero. The electric field for a given potential is then:

$$\varepsilon(\phi) = \sqrt{2\int_0^{\phi}\frac{-qNd}{\varepsilon_s}\left(1 - \exp\left(\frac{q\phi}{kT}\right)\right)d\phi} = \sin n(\phi)\frac{V_t}{L_D}\sqrt{2\left[\exp\left(\frac{\phi}{V_t}\right) - \left(\frac{\phi}{V_t}\right) - 1\right]}$$

Where the sign function equals +1 or -1 depending on the sign of ϕ and L_D is the Debye length given by, Equation is plotted in Figure using normalized parameters. Depletion occurs for negative potentials while accumulation occurs for positive potentials.

Applying Gauss's law ($\Theta = \varepsilon_s E$), we then find the relation between the total charge in the semiconductor region and the total potential across the semiconductor. The capacitance can also be obtained from:

$$C = \left|\frac{dQ}{dV_a}\right| = \varepsilon_s \left|\frac{d\mathcal{E}}{dV_a}\right| = \frac{\varepsilon_s}{L_D}\left|\frac{\exp(\frac{\phi}{V_t}) - 1}{\sqrt{2[\exp(\frac{\phi}{V_t}) - (\frac{\phi}{V_t}) - 1]}}\right|$$

where ϕ_s is the potential across the semiconductor and equals $-\phi_I + V_a$. This expression can be approximated for $\phi_s < 0$ and $|\phi_s| >> V_t$,

$$C = \frac{\varepsilon_s}{L_D}\frac{1}{\sqrt{2\frac{\phi_i - V_a - V_t}{V_t}}} = \sqrt{\frac{\varepsilon_s q N_d}{2(\phi_i - V_a - V_t)}}$$

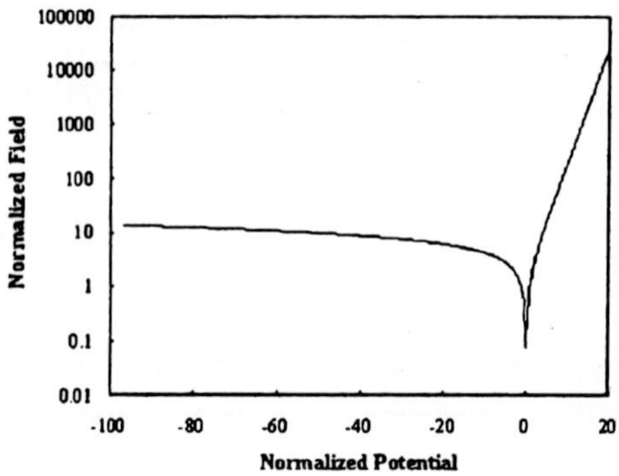

Fig. Absolute value of the normalized electric field, $|E| L_D/V_t$, versus normalized potential, ϕ/V_t

This expression equals as derived using the full depletion approximation, except for the added term, V_t, in the denominator. This expression yields the capacitance value with a relative accuracy better than 0.3 per cent for $V_a < \phi_i - 6V_t$.

Numeric Solution

A numeric solution can be obtained by integrating equation. The solution to the energy band diagram, the charge density, the electric field and the potential are shown in the figures below: Integration was started four Debye lengths to the right of the edge of the depletion region as obtained using the full depletion approximation. Initial conditions were obtained by assuming the potential at the starting point to be adequately expressed by a solution to the homogenous equation:

$$\phi = V_t \exp\left(-\frac{(x - x_d)}{L_D} \right)$$

Shown are solutions for a gold-silicon M-S junction with $\Phi_M = 4.75V$, $\chi = 4.05V$, $N_d = 10^{16}$ cm^{-3} and $\varepsilon_s/\varepsilon_0 = 11.9$.

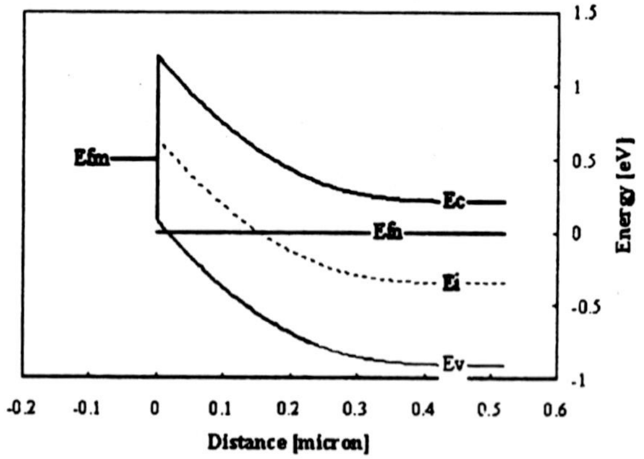

Fig. Energy band diagram of an M-S junction

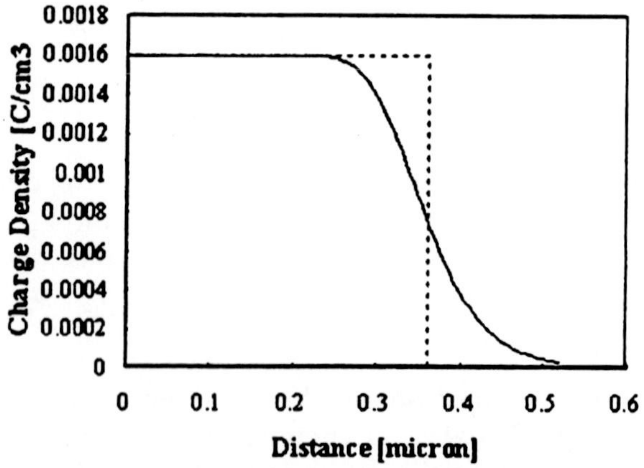

Fig. Charge density versus position in a M-S junction. The solid line is the numeric solution, and the dotted line is the solution based on the full depletion approximation.

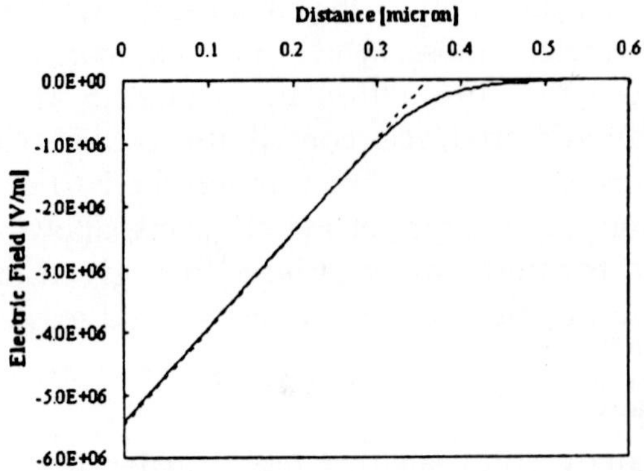

Fig. Electric field versus distance in a M-S junction. The solid line is the numeric solution, and the dotted line is the solution based on the full depletion approximation.

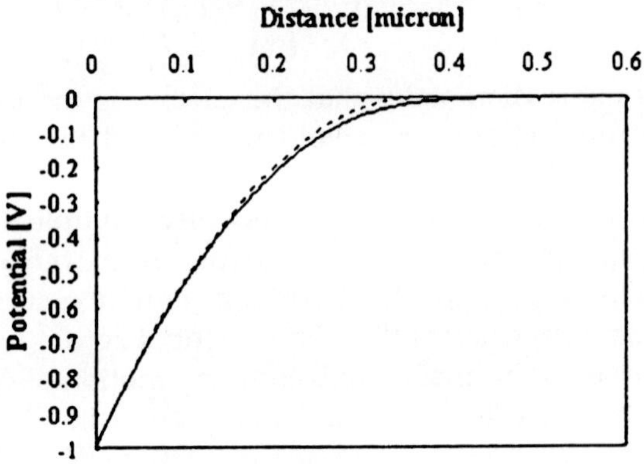

Fig. Potential versus distance of an M-S junction. The solid line is the numeric solution, and the dotted line is the solution based on the full depletion approximation.

SCHOTTKY DIODE CURRENT

The current across a metal-semiconductor junction is mainly due to majority carriers. Three distinctly different mechanisms exist: diffusion of carriers from the semiconductor into the metal, thermionic emission of carriers across the Schottky barrier and quantum-mechanical tunneling through the barrier. The *diffusion theory* assumes that the driving force is distributed over the length of the depletion layer. The *thermionic emission theory* on the other hand postulates that only energetic carriers, those, which have an energy equal to or larger than the conduction band energy at the metal-semiconductor interface, contribute to the current flow. *Quantum-mechanical tunneling* through the barrier takes into account the wave-nature of the electrons, allowing them to penetrate through thin barriers. In a given junction, a combination of all three mechanisms could exist. However, typically one finds that only one current mechanism dominates.

The analysis reveals that the diffusion and thermionic emission currents can be written in the following form:

$$J_n = qvN_c \exp\left(-\frac{\phi_B}{V_t}\right)\left(\exp\left(\frac{V_a}{V_t}\right) - 1\right)$$

This expression states that the current is the product of the electronic charge, q, a velocity, v, and the density of available carriers in the semiconductor located next to the interface. The velocity equals the mobility multiplied with the field at the interface for the diffusion current and the Richardson velocity for the thermionic emission current. The minus one term ensures that the current is zero if no voltage is applied as in thermal equilibrium any motion of carriers is balanced by a motion of carriers in the opposite direction.

The tunneling current is of a similar form, namely:

$$J_n = qv_R n \Theta$$

where v_R is the Richardson velocity and n is the density of carriers in the semiconductor. The tunneling probability term, Θ, is added since the total current depends on the carrier flux arriving at the tunnel barrier multiplied with the probability, Θ, that they tunnel through the barrier.

DIFFUSION CURRENT

This analysis assumes that the depletion layer is large compared to the mean free path, so that the concepts of drift and diffusion are valid. The resulting current density equals:

$$J_n = \frac{q^2 D_n N_c}{V_t}\sqrt{\frac{2q(\phi_i - V_a)N_d}{\varepsilon_s}}\exp(-\frac{\phi_B}{V_t})[\exp(\frac{V_a}{V_t})-1]$$

The current therefore depends exponentially on the applied voltage, V_a, and the barrier height, ϕ_B. The prefactor can more easily be understood if one rewrites it as a function of the electric field at the metal-semiconductor interface, ε_{max}:

$$\varepsilon_{max} = \sqrt{\frac{2q(\phi_i - V_a)N_d}{\varepsilon_s}}$$

yielding:

$$J_n = q\mu_n\varepsilon_{max}N_c \exp\left(-\frac{\phi_B}{V_t}\right)\left[\exp\left(\frac{V_a}{V_t}\right)-1\right]$$

so that the prefactor equals the drift current at the metal-semiconductor interface, which for zero applied voltage exactly balances the diffusion current.

THERMIONIC EMISSION

The thermionic emission theory assumes that electrons, with an energy larger than the top of the barrier, will cross the barrier provided they move towards the barrier. The actual shape of the barrier is hereby ignored. The current can be expressed as:

$$J_{MS} = A^* T^2 e^{-\phi_B/V_t}(e^{V_a/V_t} - 1)$$

where $A^* = \frac{4\pi q m^* k^2}{h^3}$ is the Richardson constant and ϕ_B is the Schottky barrier height.

The expression for the current due to thermionic emission can also be written as a function of the average velocity with which the electrons at the interface approach the barrier. This velocity is referred to as the Richardson velocity given by:

$$v_R = \sqrt{\frac{kT}{2\pi m}}$$

So that the current density becomes:

$$J_n = qv_R N_c \exp(-\frac{\phi_B}{V_t})[\exp(\frac{V_a}{V_t}) - 1]$$

TUNNELING

The tunneling current is obtained from the product of the carrier charge, velocity and density. The velocity equals the Richardson velocity, the velocity with which on average the carriers approach the barrier. The carrier density equals the density of available electrons, *n*, multiplied with the tunneling probability, Θ, yielding:

$$J_n = qvR^{n\Theta}$$

Where the tunneling probability is obtained from:

$$\Theta = \exp\left(-\frac{4}{3}\frac{\sqrt{2qm^*}}{\hbar}\frac{\phi_B^{3/2}}{\mathcal{E}}\right)$$

and the electric field equals $\varepsilon = \phi_B/L$.

The tunneling current therefore depends exponentially on the barrier height, ϕ_B, to the 3/2 power.

DERIVATION OF THE METAL-SEMICONDUCTOR JUNCTION CURRENT

Derivation of the diffusion current

We start from the expression for the total current and then integrate it over the width of the depletion region:

$$J_n = q\left(\mu_n n\varepsilon + D_n \frac{dn}{dx}\right)$$

which can be rewritten by using $\varepsilon = -d\phi/dx$ and multiplying both sides of the equation with $\exp(-\phi/V_t)$, yielding:

$$J_n \exp\left(-\frac{\phi}{V_t}\right) = qD_n\left(-\frac{n}{V_t}\frac{d\phi}{dx} + \frac{dn}{dx}\right)\exp\left(-\frac{\phi}{V_t}\right) = qD_n\frac{d}{dx}\left[n\exp\left(-\frac{\phi}{V_t}\right)\right]$$

Integration of both sides of the equation over the depletion region yields:

$$J_n = \frac{qD_n\, n \exp(-\frac{\phi}{V_t})\Big|_0^{x_d}}{\int_0^{x_d} \exp(-\frac{\phi}{V_t})\,dx} = \frac{qD_n\, N_c \exp(-\frac{\phi_B}{V_t})\left[\exp(\frac{V_a}{V_t}) - 1\right]}{\int_0^{x_d} \exp(-\frac{\phi^*}{V_t})\,dx}$$

Where the values listed in Table were used for the electron density and the potential:

x	n(x)	$\phi(x)$
0	$N_c \exp(-\phi_B/V_t)$	$-\phi_i + V_a$
x_d	$N_d = N_c \exp(-\phi_B/V_t)\exp(\phi_i/V_t)$	0

Boundary conditions used to solve equation

and $\phi^* = \phi + \phi_I - V_a$. The integral in the denominator can be solved using the potential obtained from the full depletion approximation solution, or:

$$\phi = -\frac{qN_d}{2\varepsilon_s}(x - x_d)^2$$

so that ϕ^* can be written as:

$$\phi^* = -\frac{qN_d}{\varepsilon_s}x\left(x_d - \frac{x}{2}\right) \cong \frac{qN_d}{\varepsilon_s}xx_d = (\phi_i - V_a)\frac{2x}{x_d}$$

where the second term is dropped since the linear term is dominant if x << *xsub>d*. Using this approximation one can solve the integral as:

$$\int_0^{x_d} \exp(-\frac{\phi^*}{V_t})\,dx \cong x_d \frac{V_t}{2(\phi_i - V_a)}$$

for $(\phi_i - V_a) > V_t$. This yields the final expression for the current due to diffusion:

$$J_n = \frac{qD_nN_c}{V_t}\sqrt{\frac{2q(\phi_i - V_a)N_d}{\varepsilon_s}}\exp(-\frac{\phi_B}{V_t})[\exp(\frac{V_a}{V_t}) - 1]$$

This expression indicates that the current depends exponentially on the applied voltage, V_a, and the barrier height, ϕ_B. The prefactor can be understood physically if one rewrites that term as a function of the electric field at the metal-semiconductor interface, ε_{max}:

$$\varepsilon_{max} = \sqrt{\frac{2q(\phi_i - V_a)N_d}{\varepsilon_s}}$$

yielding:

$$J_n = q\mu_n \varepsilon_{max} N_c \exp\left(-\frac{\phi B}{V_t}\right)\left[\exp\left(\frac{V_a}{V_t}\right) - 1\right]$$

so that the prefactor equals the drift current at the metal-semiconductor interface, which for zero applied voltage exactly balances the diffusion current.

Derivation of the Thermionic Emission Current

The thermionic emission theory assumes that electrons, which have an energy larger than the top of the barrier will cross the barrier, provided they move towards the barrier. The actual shape of the barrier is hereby ignored. The current can be expressed as:

$$J_{right-left} = \int_{E_c(x=\infty)+q\phi_n} qv_x \frac{dn}{dE} dE$$

For non-degenerately doped material, the density of electrons between E and $E + dE$ is given by:

$$\frac{dn}{dE} = g_c(E)F(E) = \frac{4\pi(2m^*)^{3/2}}{h^3}\sqrt{E - E_c}\exp[-(\frac{E - E_{Fn}}{kT})]$$

Assuming a parabolic conduction band (with constant effective mass m^*), the carrier energy, E, can be related to its velocity, v, by:

$$E - E_c = \frac{m^* v^2}{2}, \quad dE = m^* v dv, \text{ and } \sqrt{E - E_c} = v\sqrt{\frac{m^*}{2}}$$

Combining Equations yields:

$$\frac{dn}{dE}dE = 2\left(\frac{m^*}{h}\right)^3 \exp\left[-\frac{E_c(x=\infty) - E_{F,n}}{kT}\right]\exp\left[-\frac{m^*v^2}{2kT}\right]4\pi v^2 dv$$

when replacing v^2 by $v_x^{\ 2} + v_y^{\ 2} + v_z^{\ 2}$ and $4\Pi\ v^2 dv$ by $dv_x dv_y dv_z$ the current becomes:

$$J_{r\to l} = 2\left(\frac{m^*}{h}\right)^3 \int_{-\infty}^{\infty} \exp\left[-\frac{m^*v_y^2}{2kT}\right]dv_y \int_{-\infty}^{\infty} \exp\left[-\frac{m^*v_z^2}{2kT}\right]dv_z$$

$$\int_{-\infty}^{-v_{0x}} qv_x \exp\left[-\frac{m^*v_x^2}{2kT}\right]dv_x \exp\left[-\frac{E_c(x=0) - E_{F,n}}{kT}\right]$$

$$= 2q\left(\frac{m^*}{h}\right)^3 \frac{2\pi kT}{m^*}\exp\left[-\frac{E_c(x=0) - E_{F,n}}{kT}\right]\exp\left[-\frac{m^*v_{0x}^2}{2kT}\right]\frac{kT}{m^*}$$

using

$$\int_{-\infty}^{\infty} \exp\left[-\frac{m^*v_y^2}{2kT}\right]dv_y = \int_{-\infty}^{\infty} \exp\left[-\frac{m^*v_z^2}{2kT}\right]dv_z = \sqrt{\frac{2\pi kT}{m^*}}$$

The velocity v_{ox} is obtained by setting the kinetic energy equal to the potential across the n-type region:

$$\frac{m^* V_0^2 x}{2} = q\phi_n$$

so that v_{ox} is the minimal velocity of an electron in the quasi-neutral n-type region, needed to cross the barrier. Using

$$\phi_n = \phi_i - V_a = \phi_B - \frac{1}{q}\left[E_c(x=\infty) - E_F, n\right] - V_a$$

which is valid for a metal-semiconductor junction, one obtains:

$$J_{MS} = A^* T^2 \exp\left(-\frac{\phi B}{V_t}\right)\left[\exp\left(\frac{V_a}{V_t}\right) - 1\right]$$

where $A^* = \frac{4\pi q m^* k^2}{h^3}$ is the Richardson constant and f_B is the Schottky barrier height which equals the difference between the Fermi level in the metal, $E_{F,M}$ and the conduction band edge, E_c, evaluated at the interface between the metal and the semiconductor. The -1 term is added to account for the current flowing from right to left. The current flow from right to left is independent of the applied voltage since the barrier is independent of the band bending in the semiconductor and equal to ϕ_B. Therefore it can be evaluated at any voltage. For $V_a = 0$ the total current must be zero, yielding the -1 term.

The expression for the current due to thermionic emission can also be written as a function of the average velocity with which the electrons at the interface approach the barrier. This velocity is referred to as the Richardson velocity given by:

$$v_R = \sqrt{\frac{kT}{2\pi m}}$$

So that the current density becomes:

$$J_n = q v_R N_c \exp\left(-\frac{\phi B}{V_t}\right)\left[\exp\left(\frac{V_a}{V_t}\right)-1\right]$$

Derivation of the Tunneling Current

To derive the tunnel current, we start from the time independent Schrödinger equation:

$$-\frac{\hbar^2}{2m^*}\frac{d^2\Psi}{dx^2} + V(x)\Psi = E\Psi$$

which can be rewritten as

$$\frac{d^2\Psi}{dx^2} = \frac{2m^*(V-E)}{\hbar^2}\Psi$$

Assuming that $V(x) - E$ is independent of position in a section between x and $x+dx$ this equation can be solved yielding:

$$\Psi(x + dx) = \Psi(x)\exp(-k\,dx) \text{ with } k = \frac{\sqrt{2m^*[V(x) - E]}}{\hbar}$$

The minus sign is chosen since we assume that the particle moves from left to right. For a slowly varying potential the amplitude of the wave function at $x = L$ can be related to the wave function at $x = 0$:

$$\Psi(L) = \Psi(0)\exp\left(-\int_0^L \frac{\sqrt{2m^*[V(x) - E]}}{\hbar}dx\right)$$

This equation is referred to as the WKB approximation. From this the tunneling probability, Θ, can be calculated for a triangular barrier for which $V(x)-E = q\phi_B\,(1- x/L)$

$$\Theta = \frac{\Psi(L)\Psi^*(L)}{\Psi(0)\Psi^*(0)} = \exp\left(-2\int_0^L \frac{\sqrt{2m^*}}{\hbar}\sqrt{q\,\phi_B(1-\frac{x}{L})}\,dx\right)$$

The tunneling probability then becomes:

$$\Theta = \exp\left(-\frac{4}{3}\frac{\sqrt{2qm^*}}{\hbar}\frac{\phi_B{}^{3/2}}{\mathcal{E}}\right)$$

where the electric field equals $\varepsilon = \phi_B/L$.

The tunneling current is obtained from the product of the carrier charge, velocity and density. The velocity equals the Richardson velocity, the velocity with which on average the carriers approach the barrier while the carrier density equals the density of available electrons multiplied with the tunneling probability, yielding:

$$J_n = qV_R n\Theta$$

The tunneling current therefore depends exponentially on the barrier height to the 3/2 power.

METAL-SEMICONDUCTOR CONTACTS

Metal-semiconductor contacts are an obvious component of any semiconductor device. At the same time, such contacts cannot be assumed to have a resistance as low as that of two connected metals. In particular, a large mismatch between the Fermi energy of the metal and semiconductor can result is a high-resistance rectifying contact. A proper choice of materials can provide a low resistance Ohmic contact. However for a lot of semiconductors there is no appropriate metal available. Instead one then creates a tunnel contact. Such contact consists of a thin barrier – obtained by heavily-doping the semiconductor – through which carriers can readily tunnel. Thin interfacial layers also affect contact formation. Most metal-semiconductor contacts are annealed or alloyed after the initial deposition of the metal in an effort to further improve the contact resistivity. This section describes each of these contacts as well as an analysis of the contact resistance between a metal and a thin semiconductor layer.

OHMIC CONTACTS

A metal-semiconductor junction results in an Ohmic contact (i.e. a contact with voltage independent resistance) if the Schottky barrier height, ϕ_B, is zero or negative. In such case, the carriers are free to flow in or out of the semiconductor so that there is a minimal resistance across the contact. For an *n*-type semiconductor, this means that the workfunction of the metal must be close to or smaller than the electron affinity of the semiconductor. For a

p-type semiconductor, it requires that the workfunction of the metal must be close to or larger than the sum of the electron affinity and the bandgap energy. Since the workfunction of most metals is less than 5 V and a typical electron affinity is about 4 V, it can be problematic to find a metal that provides an Ohmic contact to *p*-type semiconductors with a large bandgap such as GaN or SiC.

TUNNEL CONTACTS

An alternate and more practical contact is a tunnel contact. Such contacts do have a positive barrier at the metal-semiconductor interface, but also have a high enough doping in the semiconductor that there is only a thin barrier separating the metal from the semiconductor. If the width of the depletion region at the metal-semiconductor interface is very thin, on the order of 3 nm or less, carriers can readily tunnel across such barrier. The required doping density for such contact is 10^{19} cm^{-3} or higher.

ANNEALED AND ALLOYED CONTACTS

The fabrication of Ohmic contacts frequently includes a high temperature step so that the deposited metals can either alloy with the semiconductor or the high-temperature anneal reduces the unintentional barrier at the interface.

In the case of silicon, one can simply deposit a metal such as aluminum and obtain a reasonable Ohmic contact. However, subsequent annealing at 475°C in a reducing ambient such as forming gas (20:1 N_2/H_2) will further improve the contact resistivity. The temperature is chosen below the eutectic temperature of the Si/Al eutectic composition. Annealing at higher temperature causes the formation of Si/Al alloys, which in turn causes pits in the silicon. This effect is also referred to as spiking and when penetrating through an underlying *p-n* junction these "spikes" dramatically affect the quality of the *p-n* junction as can be observed in the form of an enhanced leakage current or reduced breakdown voltage. The use of a reducing atmosphere avoids any further oxidation of the metal during annealing, while it can also reduce any interfacial oxide between the metal and semiconductor. Aluminum deposited onto low-doped silicon (< 10^{15} cm^{-3}) tends to form Schottky barriers, so that it is advantageous to provided a more-highly doped contact region underneath the contact metal. The small barrier height can be overcome through thermionic emission, while the contact resistance is further improved by creating a tunnel barrier using degenerately doped contact layers.

Contacts to compound semiconductors require some more attention. Selecting a material with the right workfunction might still not result in the expected Ohmic contact. This is caused by pinning of the Fermi energy at the interface due to the large number of surface states at the metal-semiconductor interface. This only leaves the tunnel contact as a viable low resistance contact. To further improve the tunnel contact one adds dopants such as germanium in the case of an *n*-type contact and zinc in the case of a *p*-type contact to the metal. An anneal at 400°C in a forming gas ambient for ten minutes causes the dopants to alloy with the semiconductor, thereby forming a thin high-doped region as desired for a tunnel contact.

CONTACT RESISTANCE TO A THIN SEMICONDUCTOR LAYER

The contact between a metal and a thin semiconductor layer can be described with the resistive network shown in Figure. This equivalent circuit is obtained by slicing the structure into small sections with length Δx, so that the contact resistance, R_1, and the semiconductor resistance, R_2, are given by:

$$R_1 = \frac{\rho_c}{W \Delta_x}$$

and

$$R_2 = R_s \frac{\Delta_x}{W}$$

where ρ_c is the contact resistance of the metal-to-semiconductor interface per unit area with units of Ωcm^2, R_s is the sheet resistance of the semiconductor layer with units of Ohms per square, and W is the width of the contact.

Using Kirchoff's laws one obtains the following relations between the voltage, $V(x)$, across the M-S interface and the current, $I(x)$, parallel to the interface at x and $x + \Delta x$.

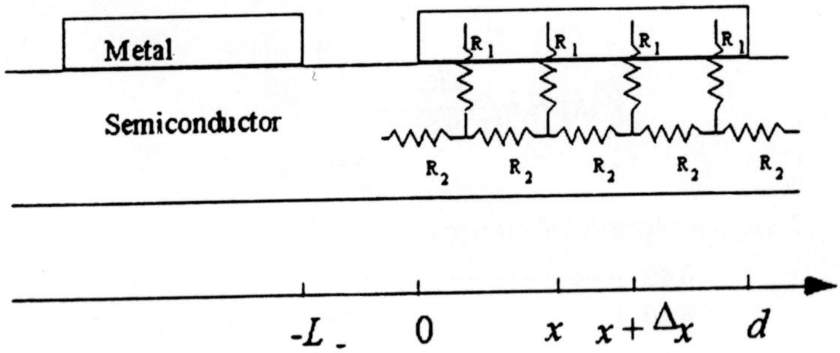

Fig. Distributed resistance model of a contact to a thin semiconductor layer.

$$V(x+\Delta x) - V(x) = I(x)R_2 = I(x)\frac{R_s}{W}\Delta x$$

$$I(x+\Delta x) - I(x) = \frac{V(x)}{R_1} = V(x)\frac{W}{\rho_c}\Delta x$$

By letting Δx approach zero one finds the following differential equations for the current, $I(x)$, and voltage, $V(x)$:

$$\frac{dV}{dx} = \frac{I(x)R_s}{W}$$

$$\frac{dI}{dx} = \frac{V(x)W}{\rho_c}$$

These equations can be combined into:

$$\frac{d^2I(x)}{dx^2} = I(x)\frac{R_s}{\rho_c} = \frac{I(x)}{\lambda^2} \text{ with } \lambda = \sqrt{\frac{\rho_c}{R_s}}$$

The parameter λ is the characteristic distance over which the current changes under the metal contact and is also referred to as the penetration length. The general solution for $I(x)$ and $V(x)$ are:

$$I(x) = I_0 \frac{\sinh \frac{d-x}{\lambda}}{\sinh \frac{d}{\lambda}}$$

$$V(x) = I_0 \frac{\lambda R_s}{W} \frac{\cosh \frac{d-x}{\lambda}}{\sinh \frac{d}{\lambda}}$$

Both are plotted in Figure:

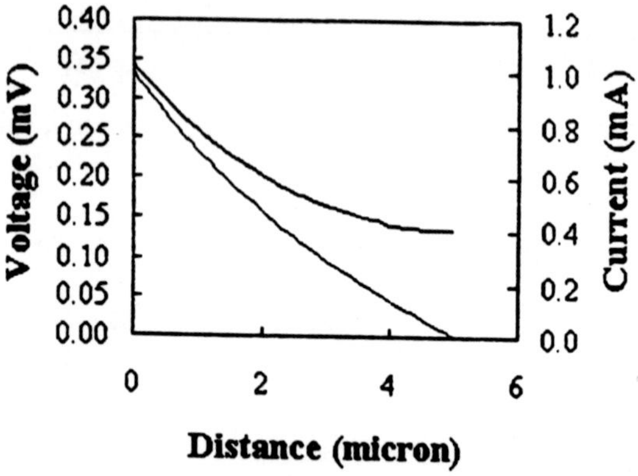

Fig. Lateral current and voltage underneath a 5 μm long and 1 mm wide metal contact with a contact resistivity of 10^{-5} Ω-cm² on a thin semiconductor layer with a sheet resistance of 100 Ω.

The total resistance of the contact is:

$$R_c = \frac{V(0)}{I(0)} = \frac{\lambda R_s}{W} \coth \frac{d}{\lambda} = \frac{\sqrt{\rho_c R_S}}{W} \coth \frac{d}{\lambda}$$

In the limit for an infinitely long contact (or $d >> \lambda$) the contact resistance is given by:

$$R_c = \frac{\sqrt{\rho_c R_S}}{W}, \text{for } d >> \lambda$$

A measurement of the resistance between a set of contacts with a variable distance L between the contacts (also referred to as a TLM structure) can therefore be fitted to the following straight line:

$$R_c = 2\frac{\sqrt{\rho_c R_S}}{W} + R_S \frac{L}{W}$$

so that the resistance per square, R_s, can be obtained from the slope, while the contact resistivity, ρ_c, can be obtained from the intersection with the y-axis. The penetration depth, λ, can be obtained from the intersection with the x-axis. This is illustrated with Figure.

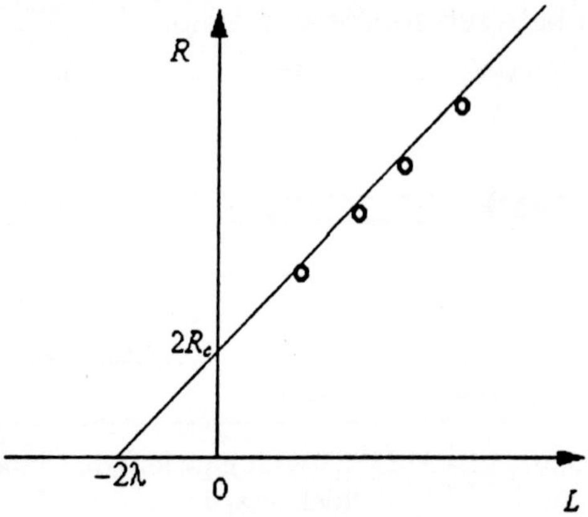

Fig. Resistance versus contact spacing, L, of a TLM structure.

In the limit for a short contact (or $d << \lambda$), the contact resistance can be approximated by expanding the hyperbolic cotangent:

$$R_c = \frac{\lambda R_s}{W}\left(\frac{\lambda}{d} + \frac{d}{3\lambda} + \ldots\right) = \frac{\rho_c}{Wd} + \frac{1}{3}R_s\frac{d}{W}\text{, for } d << \lambda$$

The total resistance of a short contact therefore equals the resistance between the contact metal and the semiconductor layer (i.e. the parallel connection of all the resistors, R_1, in Figure, plus one third of the end-to-end resistance of the conducting layer underneath the contact metal (i.e. the series connection of all resistors, R_2, in Figure.

METAL-SEMICONDUCTOR FIELD EFFECT TRANSISTOR (MESFETS)

The Metal-Semiconductor-Field-Effect-Transistor (MESFET) consists of a conducting channel positioned between a source and drain contact region as shown in the Figure. The carrier flow from source to drain is controlled by a Schottky metal gate. The control of the channel is obtained by varying the depletion layer width underneath the metal contact which modulates the thickness of the conducting channel and thereby the current between source and drain.

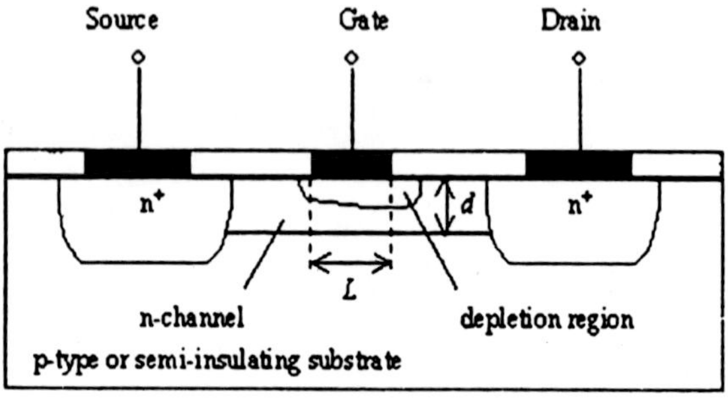

Fig. Structure of a MESFET with gate length, *L*, and channel thickness, *d*.

The key advantage of the MESFET is the higher mobility of the carriers in the channel as compared to the MOSFET. Since the carriers located in the inversion layer of a MOSFET have a wavefunction, which extends into the oxide, their mobility - also referred to as surface mobility - is less than half of the mobility of bulk material. As the depletion region separates the carriers from the surface their mobility is close to that of bulk material. The higher mobility leads to a higher current, transconductance and transit frequency of the device.

The disadvantage of the MESFET structure is the presence of the Schottky metal gate. It limits the forward bias voltage on the gate to the turn-on voltage of the Schottky diode. This turn-on voltage is typically 0.7 V for GaAs Schottky diodes. The threshold voltage therefore must be lower than this turn-on voltage. As a result it is more difficult to fabricate circuits containing a large number of enhancement-mode MESFET.

The higher transit frequency of the MESFET makes it particularly of interest for microwave circuits. While the advantage of the MESFET provides a superior microwave amplifier or circuit, the limitation by the diode turn-on is easily tolerated. Typically depletion-mode devices are used since they provide a larger current and larger transconductance and the circuits contain only a few transistors, so that threshold control is not a limiting factor. The buried channel also yields a better noise performance as trapping and release of carriers into and from surface states and defects is eliminated.

The use of GaAs rather than silicon MESFETs provides two more significant advantages: first, the electron mobility at room temperature is more than 5 times larger, while the peak electron velocity is about twice that of silicon. Second, it is possible to fabricate semi-insulating (SI) GaAs substrates, which eliminates the problem of absorbing microwave power in the substrate due to free carrier absorption.

The threshold voltage, V_T, of a MESFET is the voltage required to fully deplete the doped channel layer. This threshold voltage equals:

$$V_T = \phi_i - \frac{qN_d d^2}{2\varepsilon_s}$$

where ϕ_i is the built-in potential and d is the thickness of the doped region. This threshold voltage can also be written as a function of the pinch-off voltage V:

$$V_T = \phi_i - V_p$$

Where the pinch-off voltage equals:

$$V_P = \frac{qN_d d^2}{3\varepsilon_s}$$

The derivation of the current in a MESFET starts by considering a small section of the device between y and y + dy. The current density at that point can be written as a function of the gradient of the channel voltage:

$$J = qnv = qN_d\mu_n\varepsilon = -qN_d\mu_n\frac{dV_c(y)}{dy}$$

The drain current is related to the current density and the part of the MESFET channel that is not depleted.

$$I_D = -JW\left(d - x_n(y)\right)$$

Where the depletion layer width at position y is related to the channel voltage, $V(y)$, by:

$$x_n(y) = \sqrt{\frac{2\varepsilon_S\left(\phi_i - V_G + V_c(y)\right)}{qN_d}}$$

The equation for the current can now be integrated from source to drain, yielding:

$$I_D = qN_d\mu_n d\frac{W}{L}\left(V_C\Big|_0^{V_D} - \frac{(\phi_i - V_G + V_C)^{3/2}}{\sqrt{V_P}}\Bigg|_0^{V_D}\right)$$

Since the steady-state current in the device is independent of position, the left hand term equals I_D times L so that:

$$I_D = qN_d\mu_n d\frac{W}{L}\left(V_C\Big|_0^{V_D} - \frac{(\phi_i - V_G + V_C)^{3/2}}{\sqrt{V_P}}\Bigg|_0^{V_D}\right)$$

Integration results in:

$$I_D = q\mu_n N_d d\frac{W}{L}\left[V_D - \frac{2}{3}\left(\frac{(\phi_i - V_G + V_D)^{3/2}}{\sqrt{V_P}} - \frac{(\phi_i - V_G)^{3/2}}{\sqrt{V_P}}\right)\right]$$

This result is valid as long as the width of the un-depleted channel ($d - x_n(y)$) is positive, namely for:

$$V_D \le V_G - V_T$$

This condition also defines the quadratic region of a MESFET. For larger drain voltage, the current saturates and equals that at

$$V_D = V_G - V_T = V_{D,\text{sat}}$$

The corresponding current is the saturation current, $I_{D,sat}$:

$$I_{D,sat} = q\mu_n N_d d \frac{W}{L}\left[V_G - V_T - \frac{2}{3}\left(V_P - \frac{(\phi_i - V_G)^{3/2}}{\sqrt{V_P}}\right)\right]$$

An example of the resulting *I-V* characteristics is shown in

Figure. The corresponding device parameters are listed in Table

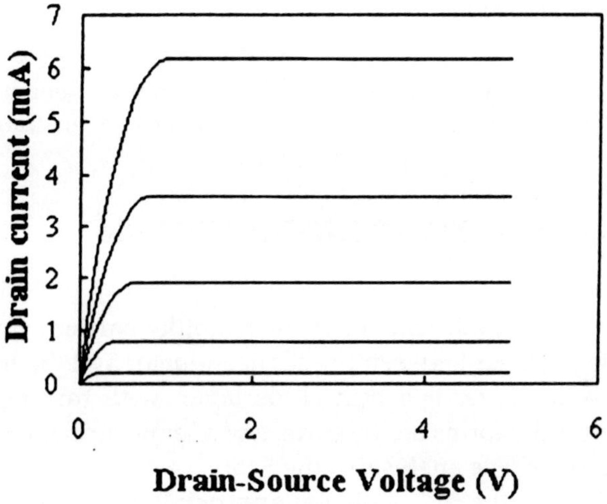

Fig. Drain current versus Drain-Source voltage at a gate-source voltage of 0.2, 0.4, 0.6 0.8 and 1.0 Volt for a silicon MESFET with built-in potential of 1 V. Channel parameters and device dimensions are listed in Table.

The transfer characteristic of a MESFET is shown in Figure and compared to a quadratic expression of the form:

$$I_{D,sat} = \mu_n \frac{\varepsilon_s}{\overline{w}} \frac{W}{L} \frac{(V_G - V_T)^2}{2}$$

where $\overline{w}$ is the average depletion layer width in the channel layer. The quadratic expression yields the same current at $V_G = \phi_i$ for $\overline{w} = 3d/8$. The close fit is at times used to justify using the simpler quadratic equation.

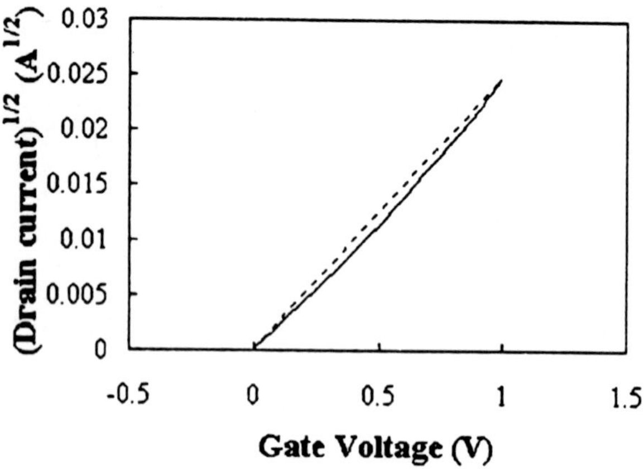

Fig. Transfer characteristic of a MESFET. Shown is the square root of the drain current of the MESFET (solid line) and a quadratic fit with $\overline{w}$=3d/8 (dotted line).

SCHOTTKY DIODE WITH AN INTERFACIAL LAYER

A more elaborate model of the Schottky barrier contains an interfacial layer between the semiconductor and the metal. Typically this layer is a thin oxide layer, with thickness *d*, which naturally forms on the surface of a semiconductor when exposed to air. The analysis of the Schottky diode can now be repeated using the full depletion approximation yielding the following relation between the total applied voltage and the depletion layer width:

$$\phi_i - V_a = \frac{qN_d x_d^2}{2\varepsilon_s} + \frac{qN_d x_d d}{\varepsilon_{ox}} = \phi_n + \phi_{ox}$$

from which the depletion layer width can be solved. The capacitance of the structure can be obtained from the series connection of the oxide and semiconductor capacitance:

$$C = \frac{1}{\dfrac{d}{\varepsilon_0 x} + \dfrac{x_d}{\varepsilon_s}} = \frac{\varepsilon_s}{L_D}\sqrt{\frac{V_t}{2\left(\phi_i^* - V_a\right)}}$$

with

$$\phi_i^* = \phi_i + \frac{qN_d d^2}{2\varepsilon_s}\left(\frac{\varepsilon_s}{\varepsilon_{ox}}\right)^2 = \phi_i + \Delta\phi$$

This expression is very similar to that of equation except that the oxide layer increases the built-in voltage. The potential ϕ_n across the semiconductor can be written as:

$$\phi_n = \phi_i - V_a + \frac{qN_d d^2}{\varepsilon_s}\left(\frac{\varepsilon_s}{\varepsilon_{ox}}\right)^2\left[1 - \sqrt{1 + \frac{2(\phi_i - V_a)}{\frac{qN_d d^2}{\varepsilon_s}\left(\frac{\varepsilon_s}{\varepsilon_{ox}}\right)^2}}\right]$$

Or alternatively,

$$\phi_n = \phi_i - V_a + 2\Delta\phi\left[1 - \sqrt{1 + \frac{\phi_i - V_a}{\Delta\phi}}\right]$$

for zero applied voltage this reduces to:

$$\phi_n = \phi_i + 2\Delta\phi\left[1 - \sqrt{1 + \frac{\phi_i}{\Delta\phi}}\right]$$

instead of simply $\phi_n = \phi_i$ when no oxide is present. This analysis can be interpreted as follows: the interfacial layer reduces the capacitance of the Schottky barrier diode, although a capacitance measurement will have the same general characteristics as an ideal Schottky barrier diode except that the built-in voltage is increased. However the potential across the semiconductor is decreased due to the voltage drop across the oxide layer, so that at low voltage the barrier for electrons flowing into the semiconductor is reduced yielding a higher current that without the oxide. This analysis assumes that the interfacial layer forms a very thin tunnel barrier, which at low voltages does not restrict the current. As the voltage applied to the Schottky barrier is more positive, the depletion layer width reduces, so that the field in the oxide also reduces and with it the voltage drop across the oxide. The current under forward bias conditions therefore approaches that of the ideal Schottky diode until the tunnel barrier restricts the current flow. This results in a higher ideality factor for Schottky barrier with an interfacial layer. From equations, we find that the effect is largest for highly doped semiconductors and interfacial layers with low dielectric constant.

The current under forward bias is then given by:

$$1 \cong I_s \exp\frac{\phi_i - \phi_n}{V_t}$$

or:

$$I = I_S \exp\frac{V_a - 2\Delta\phi\left[1 - \sqrt{1 + \frac{\phi_i - V_a}{\Delta\phi}}\right]}{V_t}$$

which, around a specific value of V_a, can be written as:

$$I = I_S^* \exp\frac{V_a}{nV_t}$$

with ideality, n:

$$n = \frac{1}{V_t \frac{dI}{dV_a}} = \frac{1}{1 - \sqrt{\frac{\Delta\phi}{\Delta\phi + \phi_i - V_a}}}$$

and saturation current I_s^*:

$$I_S^* = \frac{I(V_a = 0)}{\exp\frac{V_a}{nV_t}}$$

At $V_a = 0$ the saturation current becomes:

$$I_S^* = I(V_a = 0) = I_S \exp\frac{2\Delta\phi\left[\sqrt{1 + \phi_i/\Delta\phi} - 1\right]}{V_t}$$

As a result, an interfacial layer between the metal and semiconductor of a Schottky diode affects both the measured barrier height and built-in potential. The total potential within the device is now divided between the interfacial layer and the semiconductor. This causes the potential across the semiconductor to be lower so that carriers can more easily flow from the semiconductor into the metal, yielding a larger current. The interfacial layer also reduces the capacitance.

As an example we consider a thin 3 nm thick oxide layer at the interface of a gold-silicon Schottky diode. The energy band diagram is shown in Figure.

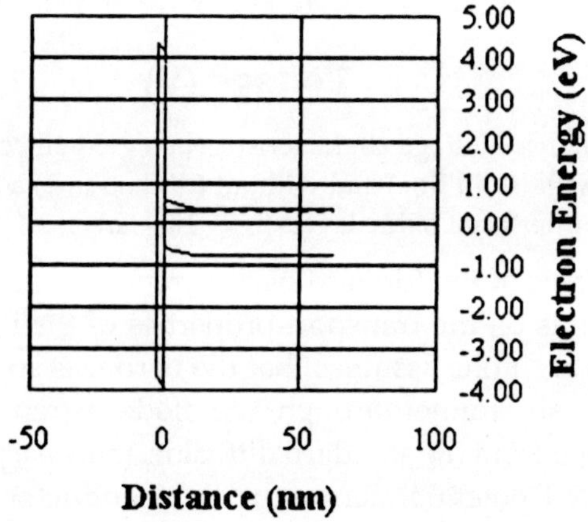

Fig. Energy band diagram of a gold-silicon M-S junction with a 3 nm interfacial oxide layer. ($V_a = 0.3$ V and $N_d = 10^{18}$ cm^{-3})

Since the interfacial layer can be viewed as an additional capacitor connected in series with the capacitance associated with the depletion layer, the total capacitance is lower than for a diode without an interfacial layer. A $1/C^2$ plot versus the applied voltage is shown in Figure.

This plot reveals that the slope remains the same, while the intercept with the voltage axis shifts to higher forward voltages. The slope remains unchanged since it depends on the doping concentration in the semiconductor, which remains unchanged. The presence of an interfacial layer therefore increases the measured built-in potential, but does not alter the extracted doping concentration.

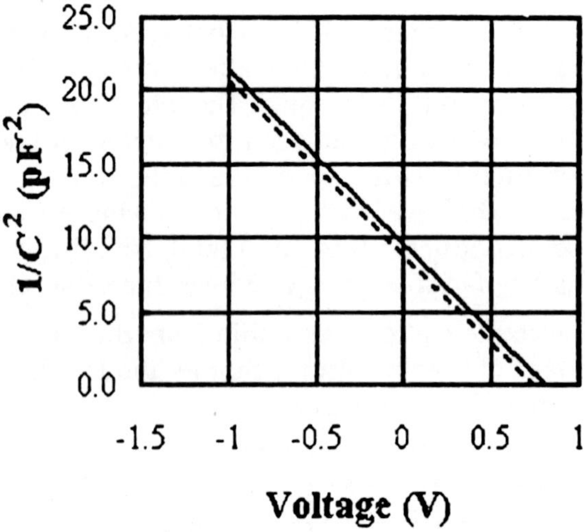

Fig. Capacitance-Voltage characteristics of a gold-silicon M-S junction with (solid line) and without (dashed line) a 3 nm interfacial oxide layer. ($N_d = 10^{18}$ cm^{-3})

The analysis of the forward bias current is more complex since it depends on the transport properties of the interfacial layer. However, if one assumes that the barrier is so thin that carriers can easily tunnel through, the diode current analysis can be obtained from the standard diffusion analysis, provided that the altered potential across the semiconductor is taken into account.

A comparison of the current through a gold-silicon junction with and without an interfacial layer is shown in Figure. The figure reveals that the interfacial layer affects both the slope and the intercept of the forward-biased current-voltage when plotted on a semi-logarithmic scale.

In summary, an interfacial layer increases the built-in potential as measured with a -*V* measurement, decreases the internal potential across the semiconductor, which increases the measured ideality factor and saturation current. It also decreases the measured barrier height as extracted from the temperature dependence of the saturation current and limits the maximum current density.

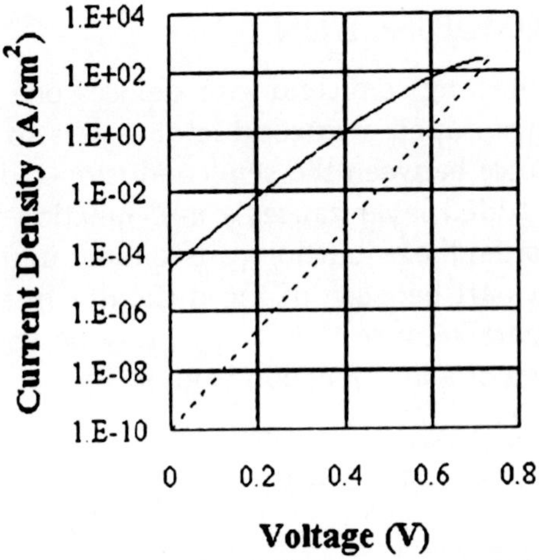

Fig. Current-Voltage characteristics of a gold-silicon M-S junction with (solid line) and without (dashed line) a 3 nm interfacial oxide layer. ($I_s = 10^{-10}$ A and $N_d = 10^{18}$ cm^{-3})

OTHER UNIPOLAR JUNCTIONS

The metal-semiconductor junction is the most studied unipolar junction, be not the only one that occurs in semiconductor devices. Two other unipolar junctions are the n-n^+ homojunction and the n-n^+ Heterojunction.

The n-n^+ homojunction frequently occurs in semiconductor devices are heavily doped regions are commonly added to reduce the overall resistance and improve the contact resistivity.We present the electrostatic analysis of the n-n^+ here in part for completeness but also to set the stage for the analysis of the n-n^+ heterojunction.

The n-n^+ heterojunction frequently occurs in heterojunction devices. Such occurrence is not always deliberate, but their analysis is , albeit complex, need when optimizing a Heterojunction device design.

In this section we present the electrostatic analysis of the n-n^+ homojunction and heterojunction as well as the analysis of the n-n^+ heterojunction current.

THE *N*-*N*$^+$ HOMOJUNCTION

When contacting semiconductor devices one very often includes highly doped semiconductor layers to lower the contact resistance between the semiconductor and the metal contact. This added layer causes a n-n^+ junction within the device. Most often these junctions are ignored in the analysis of devices, in part because of the difficulty treating them correctly, in part because they can simply be ignored. The build-in voltage of a n-n^+ junction is given by:

$$\phi_i = \frac{1}{q}(E_{Fn^+} - E_{Fn}) = V_t \ln \frac{N_d^+}{N_d}$$

Which means that the built-in voltage is about 59.4 meV if the doping concentrations differ by a factor 10. It is because of this small built-in voltage that this junction is often ignored. However large variations in doping concentration do cause significant potential variations.

The influence of the n-n^+ junction must be evaluated in conjunction with its current voltage characteristics: if the n-n^+ junction is in series with a p-n diode, the issue is whether or not the n-n^+ junction affects the operation of the p-n junction in any way. At low current densities one can expect the p-n diode to dominate the current flow, whereas at high current densities the n-$^+$ junction could play a role if not designed properly.

For the analysis of the n-n^+ junction we start from a flat band energy band diagram connecting the two regions in absence of an electric field. One can visualize that electrons will flow from the n^+ region and accumulate in the n-type region. However, since the carrier concentration must be continuous (this is only required in a homojunction), thecarrier density in the n-type region is smaller that the doping concentration of the $^+$ region, and the $^+$ region is not completely depleted. The full depletion approximation is therefore not applicable. Instead one recognizes the situation to be similar to that of a metal-semiconductor junction: the n^+ region is depleted but has a small voltage across the semiconductor as in a Schottky barrier with small applied voltage, whereas the n-type region is accumulated as in an Ohmic contact. A general solution of this structure requires the use of equation.

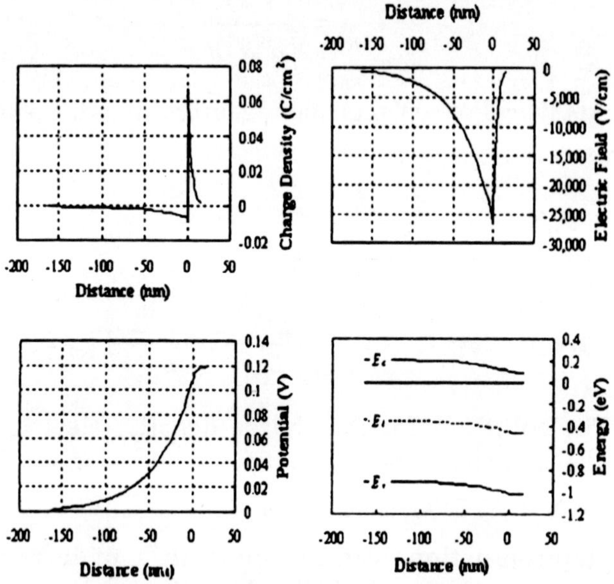

Fig. Charge, electric field, potential and energy banddiagram in a silicon n-n^+ structure with $N_d = 10^{16}$ cm^{-3}, $N_d^+ = 10^{17}$ cm^{-3} and V_a = 0.

An approximate solution can be obtained in the limit where the potential across both regions is smaller than the thermal voltage. The charge in the n-n^+ structure region is then given by solving the linearized Poisson equation:

$$\rho(x) = -\frac{\varepsilon_{sn}}{L_{Dn}} \frac{\phi_i - V_a}{\left(L_{Dn} + L_{Dn^+}\right)} \exp\frac{x}{L_{Dn}} \quad \text{for } x < 0$$

$$\rho(x) = \frac{\varepsilon_{sn^+}}{L_{Dn^+}} \frac{\phi_i - V_a}{\left(L_{Dn} + L_{Dn^+}\right)} \exp\frac{-x}{L_{Dn^+}} \quad \text{for } x > 0$$

where the n-n^+ interface is located at $x = 0$, and $L_{D,n}$ and $L_{D,n}{}^+$ are the extrinsic Debye lengths in the material, given by:

$$L_{Dn} = \sqrt{\frac{\varepsilon_s kT}{q^2 Nd}}$$

$$L_{Dn^+} = \sqrt{\frac{\varepsilon_s kT}{q^2 N^+ d}}$$

Applying Poisson's equation again one finds the potentials to be:

$$\phi_n(x) = L_{Dn} \frac{\phi_i - V_a}{L_{Dn} + L_{Dn^+}} \exp \frac{x}{L_{Dn}} \quad \text{for } x < 0$$

$$\phi_{n^+}(x) = (\phi_i - V_a)[1 - \frac{L_{Dn^+}}{L_{Dn} + L_{Dn^+}} \exp \frac{-x}{L_{Dn^+}}] \quad \text{for } x < 0$$

The solutions for the charge density, electric field, potential and energy band diagram are plotted in the Figure:

THE *N-N*⁺ HETEROJUNCTION

Heterojunctions can be found in a wide range of heterojunction devices including laser diodes, high electron mobility transistors (HEMTs) and heterojunction bipolar transistors (HBTs). Of those, the HEMT naturally contains such heterojunction, while the other devices could contain an unintentional n-n^+ heterojunction.

As a starting point of the analysis, consider a n-n^+ heterojunction including a spacer layer with thickness d as shown in Figure.

The built-in voltage for a n-n^+ heterojunction with doping concentrations N_d and N_d^+ is given by:

$$\phi_i = \frac{1}{q}(E_{Fn^+} - E_{Fn}) = V_t \ln \frac{N_d^+ N_{c,n}}{N_d N_{c,n^+}} + \Delta E_c$$

Where $N_{c,n}$ and $N_{c,n}^+$ are the effective densities of states of the low and high-doped region respectively. Unlike a homojunction, the heterojunction can have a built-in voltage, which is substantially larger than the thermal voltage.

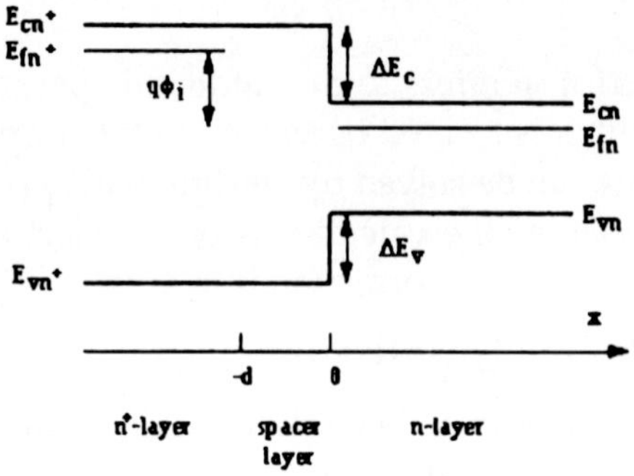

Fig. Flatband energy diagram of a n-n^+ heterojunction including a spacer layer with thickness d.

This justifies using the full depletion approximation for the depleted region. For the accumulated region one also has to consider the influence of quantization of energy levels as carriers are confined by the electric field and the hetero-interface. In the next two sections we analyze the energy band diagram of the n-$^+$ heterojunction with and without the inclusion of quantization and compare the two solutions.

Analysis Without Quantization

For the classic case where the material does not become degenerate at the interface one can use to find the total charge in the accumulation layer:

$$-Q_{acc} = \varepsilon_{s,n}\mathcal{E} = \varepsilon_{s,n}\frac{V_t}{L_D}\sqrt{2\left[\exp\frac{\phi_n}{V_t} - \frac{\phi_n}{V_t} - 1\right]}$$

while the potentials and the field can be solved for a given applied voltage using:

$$\phi_{n^+} + \phi_{sp} + \phi_n = \phi_i - V_a$$

$$\phi_{n^+} = \frac{\varepsilon_{s,n}^2 E_n^2}{2q\varepsilon_{s,n^+} N_{d^+}}$$

$$\phi_{sp} = \frac{\varepsilon_{sn} E_n d}{\varepsilon_{s,sp}}$$

The subscript *sp* refers to the undoped spacer layer with thickness *d*, which is located between the two doped regions. These equations can be solved by starting with a certain value of ϕ_n, which enables the calculation of the electric field, the other potentials and the corresponding voltage, V_a.

Analysis Including Quantization

The analysis of an n-n^+ heterojunction including quantized levels is more complicated because the energy levels depend on the potential, which can only be calculated if the energy levels are known. A self-consistent calculation is therefore required to obtain a correct solution. An approximate method, which also clarifies the steps needed for a correct solution is described below.

Starting from a certain density of electrons per unit area, N_s, which are present in the accumulation layer, one finds the field at the interface:

$$\varepsilon_n = \frac{qN_s}{\varepsilon_{s,n}}$$

We assume that only the $n = 1$ energy level is populated with electrons. The minimal energy can be expressed as a function of the electric field:

$$\varepsilon_{1n} = \left(\frac{\hbar^2}{2m^*}\right)^{1/3} \left(\frac{9\pi}{8} qE_n\right)^{2/3}$$

The bandgap discontinuity ΔE_c can then be related to the other potentials of the junction, yielding:

$$\Delta E_c - qV_a = E_{ln} + kTln\left[\exp\left(\frac{\varepsilon_{s,n}\varepsilon_n}{qN_{c,qw}}\right) - 1\right] + kTln\frac{N_{c,n^+}}{N_{d^+}} + q\phi_{n^+} + q\phi_{sp}$$

where the potentials, $\phi_n{}^+$ and ϕ_{sp}, in turn can be expressed as a function of ε_n:

$$\phi_{n^+} = \frac{\varepsilon_{s,n}^2 \varepsilon_n^2}{2q\varepsilon_{s,n^+} N_{d^+}}$$

$$\phi_{sp} = \frac{\varepsilon_{s,n} \varepsilon_n d}{\varepsilon_{s,sp}}$$

These equations can be combined into one transcendental equation as a function of the electric field, ε_n.

$$\Delta E_c - qV_a = E_{1n} + kT \ln[\exp(\frac{\varepsilon_{s,n}\varepsilon_n}{qN_{c,qw}}) - 1]$$

$$+ kT \ln \frac{N_{c,n^+}}{N_{d^+}} + \frac{\varepsilon_{s,n}^2 \varepsilon_n^2}{2\varepsilon_{s,n^+} N_{d^+}} + \frac{q\,\varepsilon_{s,n}\varepsilon_n d}{\varepsilon_{s,sp}}$$

Once ε_n is known all potentials can be obtained.

Comparison of the Two Solutions with and without Quantization

We now compare both approaches by presenting a numerical solution obtained by implementing the equations above for the case with as well as that without quantization. Keep in mind that neither is exact as it would require a self-consistent numerical analysis that includes the calculation of the potential based on the quantum mechanical distribution of the charge in each of the quantized levels.

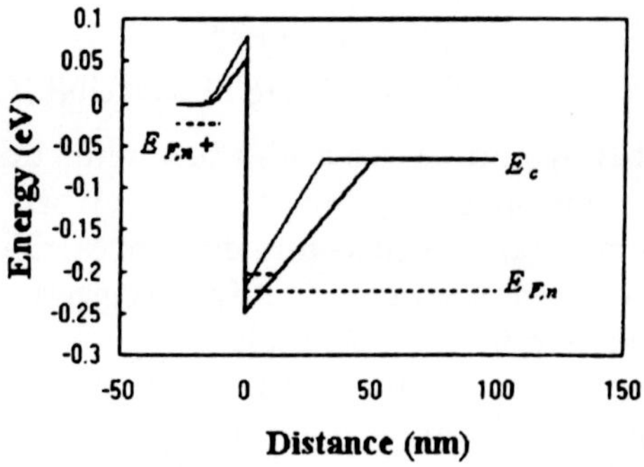

Fig. Energy banddiagram of a $Al_{0.4}Ga_{0.6}As$/GaAs n^+-n heterostructure with $N_d^+ = 10^{17}$ cm^{-3}, $N_d = 10^{16}$ cm^{-3}, d = 10 nm and V_a = 0.15 V.

Comparison of analysis without quantization (upper curve) to that with quantization (lower curve)

The energy band diagram and the sheet charge versus applied voltage are presented in Figures.

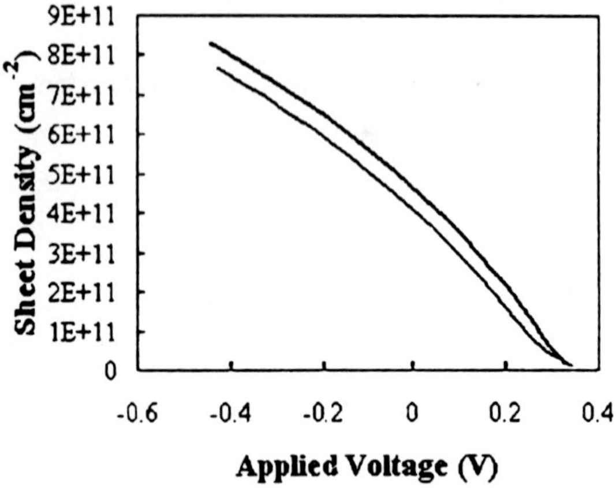

Fig. Electron density, N_s, in the accumulation region versus applied voltage, V_a, with quantization (top curve) and without quantization (bottom curve).

From the figures one finds that the analysis without quantization predicts a larger barrier to the left of the interface and a larger sheet charge for a given voltage. The actual solution is expected to be somewhere in between the two presented here, especially for the case where more than one quantized level exists and is occupied.

CURRENTS ACROSS A N^+-N HETEROJUNCTION

Current transport across a n^+-n heterojunction is similar to that of a metal-semiconductor junction: Diffusion, thermionic emission as well as tunneling of carriers across the barrier can occur. However to identify the current components one must first identify the potentials ϕ_n^+ and ϕ_n by solving the electrostatic problem. From the band diagram one finds that a barrier exists for electrons going from the n^+ to the n-doped region as well as for electrons going in the opposite direction.

The analysis in the first section discusses the thermionic emission and yields a closed form expression based on a set of specific assumptions. The derivation also illustrates how a more general expression could be obtained. The next section describes the current-voltage characteristics of carriers traversing a depletion region, while the last section discusses how both effects can be combined.

Thermionic Emission Current Across a n^+-n Heterojunction

The total current due to thermionic emission across the barrier is given by the difference of the current flowing from left to right and the current flowing from right to left. Rather than re-deriving the expression for thermionic emission, we will apply equation to the n^+-n heterojunction. One complication arises from the fact that the effective mass of the carriers is different on each side of the hetero-junction which would seem to indicate that the Richardson constant is different for carrier flow from left to right compared to the flow from right to left. A more detailed analysis reveals that the difference in effective mass causes a quantum mechanical reflection at the interface, causing carriers with the higher effective mass to be reflected back while carriers with the smaller effective mass are to first order unaffected. We therefore use equation for flow in both directions while using the Richardson constant corresponding to the smaller of the two effective masses, yielding:

$$J_{HJ} = A^* T^2 \{\exp[-\frac{E_c(x=0) - E_{F,n^+} + q\phi_{n^+}}{kT}] - \exp[-\frac{E_c(x=0) - E_{F,n} + \Delta E_c}{kT}]\}$$

where the potentials are related to the applied voltage by:

$$\phi_{n^+} + \phi_n = \phi_i - V_a$$

and the built-in voltage is given by:

$$\phi_i = \frac{\Delta E_c - (E_c(x=-\infty) - E_{Fn^+}) + (E_c(x=\infty) - E_{Fn})}{q}$$

Combining these relations yields:

$$J_{HJ} = A^*T^2 \exp[\frac{\phi_n}{V_t}]\exp[-\frac{\phi_B^*}{V_t}][\exp\frac{V_a}{V_t} - 1]$$

where the barrier height ϕ_B^* is defined as:

$$\phi_B^* = \frac{\Delta E_C + (E_C(x = \infty) - E_{Fn})}{q}$$

Assuming full depletion in the n^+ depletion region and using equation for the accumulated region, the charge balance between the depletion and accumulation layer takes the following form:

$$\sqrt{2\varepsilon_{s,n^+} qN_{d^+} \phi_{n^+}} = \varepsilon_{s,n}\sqrt{2[\exp\frac{\phi_n}{V_t} - \frac{\phi_n}{V_t} - 1]}$$

Combining equations yields a solution for ϕ_n^+ and ϕ_n.

For the special case where $\varepsilon_{sn}{}^+N_d{}^+ = \varepsilon_{sn}$I>$N_d$ and $\phi_n >> V_t$ these equations reduce to:

$$\phi_i - V_a = \phi_{n^+} + \phi_n = V_t \exp\frac{\phi_n}{V_t}$$

The current can then be expressed as a function of the applied voltage V_a

$$R_{HJ} = \frac{AJ_{HJ}}{V_a} = \frac{AA^*T^2}{V_t}\exp(-\frac{\phi_B}{V_t})$$

Whereas this expression is similar to that of a metal-semiconductor barrier, it differs in that the temperature dependence is somewhat modified and the reverse bias current increases almost linearly with voltage. Under reverse bias, the junction can be characterized as a constant resistance, R_{HJ}, which equals:

$$n \cong n - n_{p0} = n_{p0}(\exp\frac{V_a}{V_t} - 1)\exp(-\frac{x}{L_n})$$

where A is the area of the junction. This shows that the resistance changes exponentially with the barrier height. Grading of the heterojunction is typically used to reduce the spike in the energy band diagram and with it the resistance across the interface.

Calculation of the Current and quasi-Fermi Level throughout a Depletion Region

We typically assume the quasi-Fermi level to be constant throughout the depletion region. This assumption can be justified for a homojunction but is not necessarily correct for a heterojunction p-n diode.

For a homojunction p-n diode we derived the following expression for the minority carrier density in the quasi-neutral region of a "long" diode

$$n \cong n - n_{p0} = n_{p0}(\exp\frac{V_a}{V_t} - 1)\exp(-\frac{x}{L_n})$$

so that the maximum change in the quasi-Fermi level, which occurs at the edge of the depletion region, equals:

$$\frac{dF_n}{dx} = \frac{d(kT\ln\frac{n}{n_i})}{dx} \cong \frac{kT}{L_n}$$

so that the change of the quasi-Fermi level can be ignored if the depletion region width is smaller than the diffusion length as is typically the case in silicon p-n diodes.

For a hetero-junction p-n diode one can not assume that the quasi-Fermi level is continuous, especially when the minority carriers enter a narrow bandgap region in which the recombination rate is so high that the current is limited by the drift/diffusion current in the depletion region located in the wide bandgap semiconductor.

The current density can be calculated from:

$$J_n = q\mu_n n\varepsilon + qD_n\frac{dn}{dx}$$

Assuming the field to be constant throughout the depletion region one finds for a constant current density the following expression for the carrier density at the interface:

$$n = \frac{J}{q\mu_n \varepsilon} + N_d \exp\left(-\frac{\varepsilon x_n}{V_t}\right)$$

While for zero current one finds, for an arbitrary field

$$n = N_d \exp\left(-\int_{-\infty}^{0} \frac{\varepsilon(x)\,dx}{V_t}\right) = N_d \exp\left(-\frac{\phi_n}{V_t}\right)$$

Combining the two expressions we postulate the following expression for the carrier density:

$$n = \frac{J}{q\mu_n \varepsilon_{\max}} + N_d \exp\left(-\frac{\phi_n}{V_t}\right)$$

The carrier density can also be expressed as a function of the total change in the quasi-Fermi level across the depletion region, ΔE_{fn};

$$n = N_d \exp\left(-\frac{\phi_n}{V_t}\right) \exp\left(-\frac{\Delta E_{F,n}}{V_t}\right)$$

which yields the following expressions for the current density due to drift/diffusion:

$$J = -q\mu_n \varepsilon_{\max} N_d \exp\left(-\frac{\phi_n}{V_t}\right)\left(\exp\left(-\frac{\Delta E_{F,n}}{kT}\right) - 1\right)$$

where is $\varepsilon_{\max}$ the field at the heterojunction interface. If ΔE_{fn} equals the applied voltage, as is the case for an n^+-n heterostructure, this expression equals:

$$J = q\mu_n \varepsilon_{\max} N_d \exp\left(-\frac{\phi_n}{V_t}\right)\left(\exp\left(-\frac{V_a}{V_t}\right) - 1\right)$$

which reduces for a M-S junction to:

$$J_{dd} = J_{\text{thermionic}} \frac{\mu_n \varepsilon_{\max}}{v_R}$$

so that thermionic emission dominates for $v_R << \mu_n\ \varepsilon_{\max}$ or when the drift velocity is larger than the Richardson velocity.

Calculation of the Current due to Thermionic Emission and Drift/diffusion

The calculation of the current through an n^+-n junction due to thermionic emission and drift/diffusion becomes straightforward once one realises that the total applied voltage equals the sum of the quasi-Fermi level variation, ΔE_{fn}, across each region. For this analysis we therefore rewrite the current expressions as a function of ΔE_{fn}, while applying the expression for the drift/diffusion current to the n^+ material.

$$J_{\text{thermionic}} = A^* T^2 \exp\left(\frac{\phi_n}{V_t}\right)\exp\left(-\frac{\phi^* B}{V_t}\right)\left[\exp\frac{\Delta E_{F,nl}}{kT} - 1\right]$$

$$J_{drift/diffusion} = -q\mu_n + \varepsilon_{\max} N_{c,n} + \exp\left(-\frac{\phi^* B}{V_t}\right)\left[\exp\frac{\Delta E_{F,n^2}}{kT} - 1\right]$$

CURRENTS THROUGH INSULATORS

Current mechanisms through materials, which do not contain free carriers, can be distinctly different from those in doped semiconductors or metals. The following section discusses Fowler-Nordheim Tunneling, Poole-Frenkel emission, Space charge effects as well as Ballistic transport. While these current mechanisms are not related to Metal-Semiconductor junctions, they are presented here since they represent unipolar current mechanisms.

Fowler-Nordheim Tunneling

Fowler-Nordheim tunneling has been studied extensively in Metal-Oxide-Semiconductor structures where it has been shown to be the dominant current mechanism, especially for thick oxides. The basic idea is that quantum mechanical tunneling from the adjacent conductor into the insulator limits the current through the structure. Once the carriers have tunneled into the insulator they are free to move within the valence or conduction band of the insulator.The following relation between the current density, J_{FN}, and the electric field in the oxide, ε_{ox}:

$$J_{FN} = C_{FN}\mathcal{E}_{ox}^2 \exp_{ox}\left(-\frac{4}{3}\frac{\sqrt{2m_{ox}^*}}{q\hbar}\frac{(q\phi_B)^{3/2}}{\mathcal{E}_{ox}}\right)$$

where ϕ_B is the barrier height at the conductor/insulator interface.

To check for this current mechanism, experimental *I-V* characteristics are typically plotted as $\ln(J_{FN}/\varepsilon_{ox}^2)$ versus $1/\varepsilon_{ox}$, a so-called Fowler-Nordheim plot. Provided the effective mass of the insulator is known (for SiO2, $m_{ox}^* = 0.42\ m_0$) one can then fit the experimental data to a straight line yielding a value for the barrier height.

It is this type of measurement, which has yielded experimental values for the conduction band difference between different metals and silicon dioxide. It is important to note that carriers must tunnel through the insulator, which requires:

$$\varepsilon_{ox} d \geq \phi_B$$

as is typically the case for thick oxides and high electric fields.

Poole-Frenkel Emission

The expression for Fowler-Nordheim tunneling implies that carriers are free to move through the insulator. Whereas this is indeed the case in thermally grown silicon dioxide it is frequently not so in deposited insulators, which contain a high density of structural defects. Silicon nitride (Si_3N_4) is an example of such material. The structural defects cause additional energy states close to the band edge, called traps.

These traps restrict the current flow because of a capture and emission process, thereby becoming the dominant current mechanism. The current is a simple drift current described by

$$J = qn\mu\varepsilon_N$$

while the carrier density depends exponentially on the depth of the trap, which is corrected for the electric field

$$n = n_0 \exp[-\frac{q}{kT}(\phi_B - \sqrt{\frac{q\varepsilon_N}{\pi\varepsilon_N}})]$$

The total current then equals:

$$J_{PF} = qn_0\mu\varepsilon_N \exp\left[-\frac{q}{kT}\left(\phi B - \sqrt{\frac{q\varepsilon N}{\pi\varepsilon_N}}\right)\right]$$

The existence of a large density of shallow traps in CVD silicon nitride makes Poole-Frenkel emission a frequently observed and well-characterized mechanism.

Space Charge Limited Current

Both Fowler-Nordheim tunneling and Poole-Frenkel emission mechanism yield very low current densities with correspondingly low carrier densities. For structures where carriers can readily enter the insulator and freely flow through the insulator one finds that the resulting current and carrier densities are much higher. The high density of these charged carriers causes a field gradient, which limits the current density. This situation occurs in lowly doped semiconductors and vacuum tubes. We start from an expression for the drift current and Gauss's law (where we assume that the insulator contains no free carriers if no current flows)

$$J = qp\mu\varepsilon$$

$$\frac{d\varepsilon}{dx} = \frac{dp}{\varepsilon}$$

Positively charged holes are assumed in this derivation, but the result equally holds for electrons. Next we can eliminate the carrier density, p, yielding:

$$\frac{J}{\varepsilon\mu} = \varepsilon \frac{d\varepsilon}{dx}$$

Integrating this expression from 0 to x, while we assuming the electric field to be zero at $x = 0$ one obtains:

$$\frac{J}{\varepsilon\mu} = \frac{\varepsilon^2}{2} \text{ or } \varepsilon(x) = \sqrt{\frac{2xJ}{\varepsilon\mu}}$$

integrating once again from $x = 0$ to $x = d$ with $V(0) = V$ and $V(d) = 0$, one finds:

$$V = \int_0^d \mathcal{E} dx = \sqrt{\frac{2J}{\varepsilon\mu}} \frac{d^{3/2}}{3/2}$$

from which one obtains the expression for the space-charge-limited current:

$$J = \frac{9\varepsilon\mu V^2}{8d^3}$$

BALLISTIC TRANSPORT IN INSULATORS

Ballistic transport is carrier transport without scattering or any other mechanism, which would cause a loss of energy. Combining energy conservation, current continuity and Gauss's law one finds the following current-voltage relation:

$$J = \frac{4\varepsilon}{9} \sqrt{\frac{2q}{m^*}} \frac{V^{3/2}}{d^2}$$

where d is the thickness of the insulator and m^* is the effective mass of the carriers.

Chapter 3

p-n Junctions

INTRODUCTION

P-n junctions consist of two semiconductor regions of opposite type. Such junctions show a pronounced rectifying behaviour. They are also called p-n diodes in analogy with vacuum diodes.

The p-n junction is a versatile element, which can be used as a rectifier, as an isolation structure and as a voltage-dependent capacitor. In addition, they can be used as solar cells, photodiodes, light emitting diodes and even laser diodes. They are also an essential part of Metal-Oxide-Silicon Field-Effects-Transistors (MOSFETs) and Bipolar Junction Transistors (BJTs).

STRUCTURE AND PRINCIPLE OF OPERATION

A p-n junction consists of two semiconductor regions with opposite doping type as shown in Figure 4.2.1. The region on the left is *p*-type with an acceptor density N_a, while the region on the right is *n*-type with a donor density N_d. The dopants are assumed to be shallow, so that the electron (hole) density in the *n*-type (*p*-type) region is approximately equal to the donor (acceptor) density.

We will assume, unless stated otherwise, that the doped regions are *uniformly* doped and that the transition between the two regions is abrupt. We will refer to this structure as an *abrupt* p-n junction.

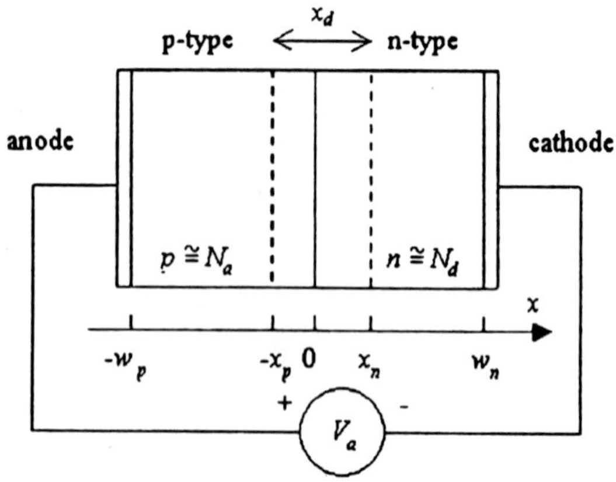

Fig. Cross-section of a p-n junction

Frequently we will deal with p-n junctions in which one side is distinctly higher-doped than the other. We will find that in such a case only the low-doped region needs to be considered, since it primarily determines the device characteristics. We will refer to such a structure as a *one-sided* abrupt p-n junction.

The junction is biased with a voltage V_a as shown in Figure. We will call the junction forward-biased if a positive voltage is applied to the *p*-doped region and reversed-biased if a negative voltage is applied to the *p*-doped region. The contact to the *p*-type region is also called the anode, while the contact to the *n*-type region is called the cathode, in reference to the *anions* or positive carriers and *cations* or negative carriers in each of these regions.

FLATBAND DIAGRAM

The principle of operation will be explained using a *gedanken* experiment, an experiment, which is in principle possible but not necessarily executable in practice. We imagine that one can bring both semiconductor regions together, aligning both the conduction and valence band energies of each region. This yields the so-called flatband diagram shown in Figure.

Note that this does not automatically align the Fermi energies, $E_{F,n}$ and $E_{F,p}$. Also, note that this flatband diagram is not an equilibrium diagram since both electrons and holes can lower their energy by crossing the junction. A motion of electrons and holes is therefore expected before thermal equilibrium is obtained.

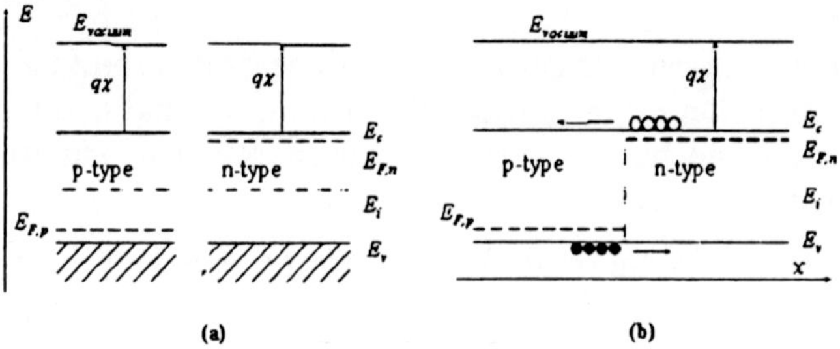

Fig. Energy band diagram of a p-n junction (a) before and (b) after merging the n-type and p-type regions

The diagram shown in Figure is called a flatband diagram. This name refers to the horizontal band edges. It also implies that there is no field and no net charge in the semiconductor.

THERMAL EQUILIBRIUM

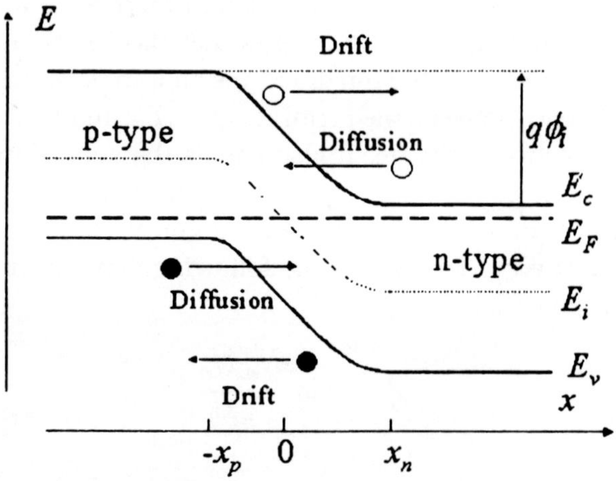

Fig. Energy band diagram of a p-n junction in thermal equilibrium

To reach thermal equilibrium, electrons/holes close to the metallurgical junction diffuse across the junction into the *p*-type/*n*-type region where hardly any electrons/holes are present. This process leaves the ionized donors (acceptors)

behind, creating a region around the junction, which is depleted of mobile carriers. We call this region the depletion region, extending from $x = -x_p$ to $x = x_n$. The charge due to the ionized donors and acceptors causes an electric field, which in turn causes a drift of carriers in the opposite direction. The diffusion of carriers continues until the drift current balances the diffusion current, thereby reaching thermal equilibrium as indicated by a constant Fermi energy. This situation is shown in Figure:

While in thermal equilibrium no external voltage is applied between the *n*-type and *p*-type material, there is an internal potential, ϕ_i, which is caused by the workfunction difference between the *n*-type and *p*-type semiconductors. This potential equals the *built-in* potential, which will be further discussed in the next section.

THE BUILT-IN POTENTIAL

The built-in potential in a semiconductor equals the potential across the depletion region in thermal equilibrium. Since thermal equilibrium implies that the Fermi energy is constant throughout the p-n diode, the built-in potential equals the difference between the Fermi energies, E_{Fn} and E_{Fp}, divided by the electronic charge. It also equals the sum of the bulk potentials of each region, ϕ_n and ϕ_p, since the bulk potential quantifies the distance between the Fermi energy and the intrinsic energy. This yields the following expression for the built-in potential.

$$\phi_i = V_t \ln \frac{N_d N_a}{n_i^2}$$

Example

An abrupt silicon p-n junction consists of a *p*-type region containing 2×10^{16} cm^{-3} acceptors and an *n*-type region containing also 10^{16} cm^{-3} acceptors in addition to 10^{17} cm^{-3} donors.

a. Calculate the thermal equilibrium density of electrons and holes in the *p*-type region as well as both densities in the *n*-type region.

b. Calculate the built-in potential of the p-n junction

c. Calculate the built-in potential of the p-n junction at 400 K.

Solution

a. The thermal equilibrium densities are:

In the *p*-type region:

$p = N_a = 2 \times 10^{16}$ cm^{-3}

$n = n_i^2/p = 10^{20}/2 \times 10^{16} = 5 \times 10^3$ cm^{-3}

In the *n*-type region

$n = N_d - N_a = 9 \times 10^{16}$ cm^{-3}

$p = n_i2/n = 10^{20}/(1 \times 10^{16}) = 1.11 \times 10^3$ cm^{-3}

b. The built-in potential is obtained from:

$$\phi_i = V_t \ln \frac{p_n n_p}{n_i^2} = 0.0259 \ln \frac{2\times10^{16} \times 9\times10^{16}}{10^{20}} = 0.79\text{ V}$$

c. Similarly, the built-in potential at 400 K equals:

$$\phi_i = V_t \ln \frac{p_n n_p}{n_i^2} = 0.0345 \ln \frac{2\times10^{16} \times 9\times10^{16}}{(4.52\times10^{12})^2} = 0.63\text{ V}$$

where the instrinsic carrier density at 400 K was obtained from example.

FORWARD AND REVERSE BIAS

We now consider a p-n diode with an applied bias voltage, V_a. A forward bias corresponds to applying a positive voltage to the anode (the *p*-type region) relative to the cathode (the *n*-type region). A reverse bias corresponds to a negative voltage applied to the cathode. Both bias modes are illustrated with Figure. The applied voltage is proportional to the difference between the Fermi energy in the *n*-type and *p*-type quasi-neutral regions.

As a negative voltage is applied, the potential across the semiconductor increases and so does the depletion layer width. As a positive voltage is applied, the potential across the semiconductor decreases and with it the depletion layer width. The total potential across the semiconductor equals the built-in potential minus the applied voltage, or:

$$\phi = \phi_i - V_a$$

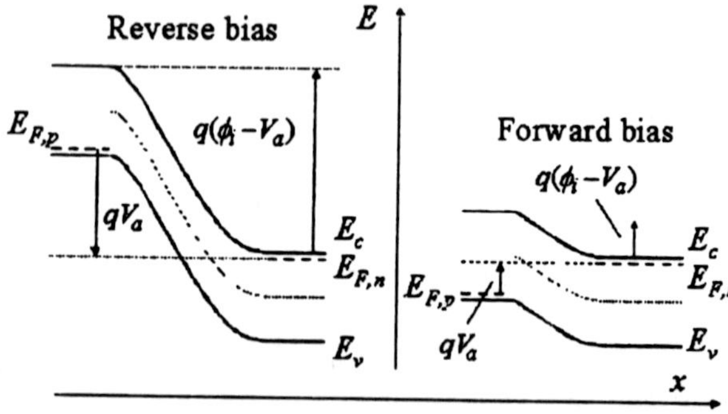

Fig. Energy band diagram of a p-n junction under reverse and forward bias

ELECTROSTATIC ANALYSIS OF A P-N DIODE

The electrostatic analysis of a p-n diode is of interest since it provides knowledge about the charge density and the electric field in the depletion region. It is also required to obtain the capacitance-voltage characteristics of the diode. The analysis is very similar to that of a metal-semiconductor junction. A key difference is that a p-n diode contains two depletion regions of opposite type.

GENERAL DISCUSSION - POISSON'S EQUATION

The general analysis starts by setting up Poisson's equation:

$$\frac{d^2\phi}{dx^2} = -\frac{\rho}{\varepsilon_s} = -\frac{q}{\varepsilon_s}\left(p - n + N_d^+ - N_a^-\right)$$

where the charge density, ρ, is written as a function of the electron density, the hole density and the donor and acceptor densities. To solve the equation, we have to express the electron and hole density, n and p, as a function of the potential, ϕ, yielding:

$$\frac{d^2\phi}{dx^2} = \frac{2qn_i}{\varepsilon_s}\left(\sinh\frac{\phi - \phi_F}{V_t} + \sinh\frac{\phi_F}{V_t}\right)$$

with

$$\sinh\frac{\phi_F}{V_t} = \frac{N_a^- - N_d^+}{2n_i}$$

where the potential is chosen to be zero in the n-type region, far away from the p-n interface.

This second-order non-linear differential equation cannot be solved analytically. Instead we will make the simplifying assumption that the depletion region is fully depleted and that the adjacent neutral regions contain no charge. This *full depletion approximation* is the topic of the next section.

THE FULL-DEPLETION APPROXIMATION

The full-depletion approximation assumes that the depletion region around the metallurgical junction has well-defined edges. It also assumes that the transition between the depleted and the quasi-neutral region is abrupt. We define the quasi-neutral region as the region adjacent to the depletion region where the electric field is small and the free carrier density is close to the net doping density.

The full-depletion approximation is justified by the fact that the carrier densities change exponentially with the position of the Fermi energy relative to the band edges. For example, as the distance between the Fermi energy and the conduction band edge is increased by 59 meV, the electron concentration at room temperature decreases to one tenth of its original value. The charge in the depletion layer is then quickly dominated by the remaining ionized impurities, yielding a constant charge density for uniformly doped regions.

We will therefore start the electrostatic analysis using an abrupt charge density profile, while introducing two unknowns, namely the depletion layer width in the *p*-type region, x_p, and the depletion region width in the *n*-type region, x_n. The sum of the two depletion layer widths in each region is the total depletion layer width x_d, or:

$$x_d = x_n + x_p$$

From the charge density, we then calculate the electric field and the potential across the depletion region. A first relationship between the two unknowns is obtained by setting the positive charge in the depletion layer equal to the negative charge. This is required since the electric field in both quasi-neutral regions must be zero. A second relationship between the two unknowns is obtained by relating the potential across the depletion layer width to the applied voltage. The combination of both relations yields a solution for x_p and x_n, from which all other parameters can be obtained.

FULL DEPLETION ANALYSIS

Once the full-depletion approximation is made, it is easy to find the charge density profile: It equals the sum of the charges due to the holes, electrons, ionized acceptors and ionized holes:

$$\rho = q\left(p - n + N_d^+ - N_a^-\right) \cong q\left(N_d^+ - N_a^-\right), \text{ for } -x_p \leq x \leq x_n$$

where it is assumed that no free carriers are present within the depletion region. For an abrupt p-n diode with doping densities, N_a and N_d, the charge density is then given by:

$$\rho(x) = 0$$

$$\rho(x) = -qN_a$$

$$\rho(x) = qN_d$$

$$\rho(x) = 0$$

This charge density, ρ, is shown in Figure.

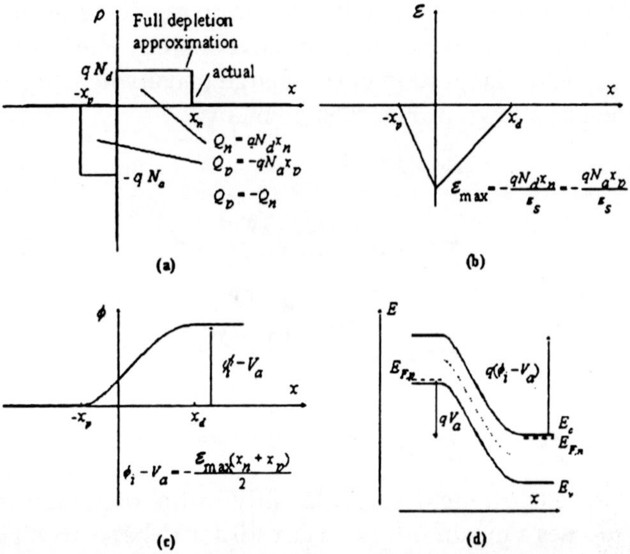

Fig. (a) Charge density in a p-n junction, (b) Electric field, (c) Potential and (d) Energy band diagram

As can be seen from Figure, the charge density is constant in each region, as dictated by the *full-depletion approximation.* The total charge per unit area in each region is also indicated on the figure. The charge in the *n*-type region, Q_n, and the charge in the *p*-type region, Q_p, are given by:

$$Q_n = qN_d x_n$$

$$Q_p = -qN_a x_n$$

The electric field is obtained from the charge density using *Gauss's law,* which states that the field gradient equals the charge density divided by the dielectric constant or:

$$\frac{d\mathcal{E}(x)}{dx} = \frac{\rho(x)}{\varepsilon_s} \cong \frac{q}{\varepsilon_s}(N_d^+(x) - N_a^-(x)), \text{ for } -x_p \leq x \leq x_n$$

The electric field is obtained by integrating equation. The boundary conditions, consistent with the full depletion approximation, are that the electric field is zero at both edges of the depletion region, namely at $x = -x_p$ and $x = x_n$. The electric field has to be zero outside the depletion region since any field would cause the free carriers to move thereby eliminating the electric field. Integration of the charge density in an abrupt p-n diode as shown in Figure is given by:

$$\mathcal{E}(x) = 0$$

$$\mathcal{E}(x) = -\frac{qN_a(x + x_p)}{\varepsilon_s}$$

$$\mathcal{E}(x) = \frac{qN_d(x - x_n)}{\varepsilon_s}$$

$$\mathcal{E}(x) = 0$$

The electric field varies linearly in the depletion region and reaches a maximum value at $x = 0$ as can be seen on Figure. This maximum field can be calculated on either side of the depletion region, yielding:

$$\varepsilon(x = 0) = -\frac{qN_a x_p}{\varepsilon_s} = -\frac{qN_d x_n}{\varepsilon_s}$$

This provides the first relationship between the two unknowns, x_p and x_n, namely:

$$N_d x_n = N_a x_p$$

This equation expresses the fact that the total positive charge in the *n*-type depletion region, Q_n, exactly balances the total negative charge in the *p*-type depletion region, Q_p. We can then combine equation with expression for the total depletion-layer width, x_d, yielding:

$$x_n = x_d \frac{N_a}{N_a + N_d}$$

and

$$x_p = x_d \frac{N_a}{N_a + N_d}$$

The potential in the semiconductor is obtained from the electric field using:

$$\frac{d\phi(x)}{dx} = -\varepsilon(x)$$

We therefore integrate the electric field yielding a piece-wise parabolic potential versus position as shown in Figure

The total potential across the semiconductor must equal the difference between the built-in potential and the applied voltage, which provides a second relation between x_p and x_n, namely:

$$\phi_i - V_a = \frac{qN_d x_n^2}{2\varepsilon_s} + \frac{qN_a x_p^2}{2\varepsilon_s}$$

The depletion layer width is obtained by substituting the expressions for x_p and x_n, into the expression for the potential across the depletion region, yielding:

$$x_d = \sqrt{\frac{2\varepsilon_s}{q}\left(\frac{1}{N_a} + \frac{1}{N_d}\right)(\phi_i - V_a)}$$

from which the solutions for the individual depletion layer widths, x_p and x_n are obtained:

$$x_n = \sqrt{\frac{2\varepsilon_s}{q}\frac{N_a}{N_d}\frac{1}{N_a + N_d}(\phi_i - V_a)}$$

$$x_p = \sqrt{\frac{2\varepsilon_s}{q}\frac{N_d}{N_a}\frac{1}{N_a + N_d}(\phi_i - V_a)}$$

Example

An abrupt silicon ($n_I = 10^{10}$ cm^{-3}) p-n junction consists of a p-type region containing 10^{16} cm^{-3} acceptors and an n-type region containing 5×10^{16} cm^{-3} donors.

a. Calculate the built-in potential of this p-n junction.

b. Calculate the total width of the depletion region if the applied voltage V_a equals 0, 0.5 and -2.5 V.

c. Calculate maximum electric field in the depletion region at 0, 0.5 and -2.5 V.

d. Calculate the potential across the depletion region in the n-type semiconductor at 0, 0.5 and -2.5 V.

Solution

The built-in potential is calculated from:

$$\phi_i = V_t \ln \frac{p_n n_p}{n_i^2} = 0.0259 \ln \frac{10^{16} \times 5 \times 10^{16}}{10^{20}} = 0.76\ \text{V}$$

The depletion layer width is obtained from:

$$x_d = \sqrt{\frac{2\varepsilon_s}{q}\left(\frac{1}{N_a} + \frac{1}{N_d}\right)(\phi_i - V_a)}$$

the electric field from

$$\varepsilon(x = 0) = -\frac{2(\phi_i - V_a)}{x_d}$$

and the potential across the n-type region equals

$$\phi_n = \frac{qN_d x_n^2}{2\varepsilon_s}$$

where

$$x_n = x_d \frac{N_a}{N_a + N_d}$$

one can also show that:

$$\phi_n = \frac{(\phi_i - V_a) N_a}{N_a + N_d}$$

This yields the following numeric values:

	$V_a = 0\text{V}$	$V_a = 0.5\text{V}$	$V_a = -2.5\text{V}$
x_d	0.315mm	0.143μm	0.703μm
ε	40kV/cm	18kV/cm	89kV/cm
ϕ_n	0.105V	0.0216V	0.522V

JUNCTION CAPACITANCE

Any variation of the charge within a p-n diode with an applied voltage variation yields a capacitance, which must be added to the circuit model of a p-n diode. This capacitance related to the depletion layer charge in a p-n diode is called the junction capacitance.

The capacitance versus applied voltage is by definition the change in charge for a change in applied voltage, or:

$$C(V_a) \stackrel{\Delta}{=} \left| \frac{dQ(V_a)}{dV_a} \right|$$

The absolute value sign is added in the definition so that either the positive or the negative charge can be used in the calculation, as they are equal in magnitude. Using equations one obtains:

$$C_j = \sqrt{\frac{q\,\varepsilon_s}{2(\phi_i - V_a)} \frac{N_a N_d}{N_a + N_d}}$$

A comparison with equation, which provides the depletion layer width, x_d, as a function of voltage, reveals that the expression for the junction capacitance, C_j, seems to be identical to that of a parallel plate capacitor, namely:

$$C_j = \frac{\varepsilon_s}{x_d}$$

The difference, however, is that the depletion layer width and hence the capacitance is voltage dependent. The parallel plate expression still applies since charge is only added at the edge of the depletion regions. The distance between the added negative and positive charge equals the depletion layer width, x_d.

The capacitance of a p-n diode is frequently expressed as a function of the zero bias capacitance, C_{j0}:

$$\frac{1}{C_j^2} = \frac{2}{q\varepsilon_s} \frac{N_a + N_d}{N_a N_d} (\phi_i - V_a)$$

Where

$$\frac{d\left(1/C_j^2\right)}{dv_a} = -\frac{2}{q\varepsilon_s}\frac{N_a+N_d}{N_aN_d}$$

A capacitance versus voltage measurement can be used to obtain the built-in voltage and the doping density of a one-sided p-n diode. When plotting the inverse of the capacitance squared, one expects a linear dependence as expressed by:

$$N_d = -\frac{2}{q\varepsilon_s}\frac{1}{\frac{d\left(1/C_j^2\right)}{dV_a}}, \quad \text{if } N_a \gg N_d$$

The capacitance-voltage characteristic and the corresponding $1/C^2$ curve are shown in Figure

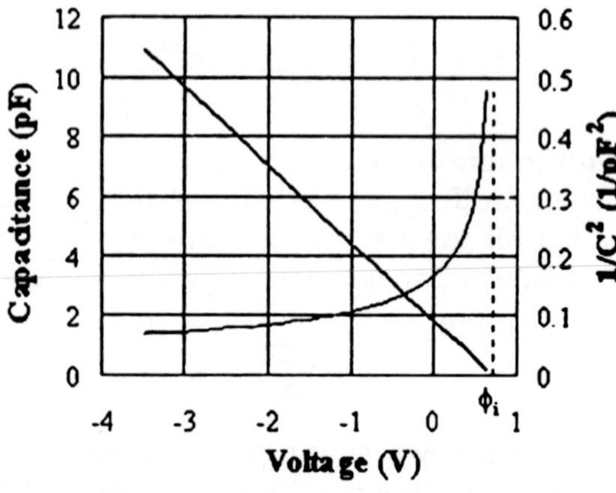

Fig. Capacitance and $1/C^2$ versus voltage of a p-n diode with $N_a = 10^{16}$ cm^{-3}, $N_d = 10^{17}$ cm^{-3} and an area of 10^{-4} cm^2.

The built-in voltage is obtained at the intersection of the $1/C^2$ curve and the horizontal axis, while the doping density is obtained from the slope of the curve.

$$\frac{d\left(1/C_j^2\right)}{dV_a} = -\frac{2}{q\varepsilon_s}\frac{N_a+N_d}{N_aN_d}$$

Example

Consider an abrupt p-n diode with $N_a = 10^{18}$ cm^{-3} and $N_d = 10^{16}$ cm^{-3}. Calculate the junction capacitance at zero bias. The diode area equals 10^{-4} cm^2. Repeat the problem while treating the diode as a one-sided diode and calculate the relative error.

Solution

The built in potential of the diode equals:

$$\phi_i = V_t \ln \frac{N_d N_a}{n_i^2} = 0.83V$$

The depletion layer width at zero bias equals:

$$x_d = \sqrt{\frac{2\varepsilon_s(\phi_i - 0)}{qN_d}} = 0.33\mu m$$

And the junction capacitance at zero bias equals:

$$C_{j0} = \left.\frac{\varepsilon_s}{x_d}\right|_{V_a=0} = 3.17\,\text{pF}$$

Repeating the analysis while treating the diode as a one-sided diode, one only has to consider the region with the lower doping density so that

$$x_d \cong x_n = \sqrt{\frac{2\varepsilon_s}{qN_d}(\phi_i - V_a)} = 0.31\mu m$$

And the junction capacitance at zero bias equals

$$C_{j0} = \left.\frac{\varepsilon_s}{x_d}\right|_{V_a=0} = 3.18\,\text{pF}$$

The relative error equals 0.5 per cent, which justifies the use of the one-sided approximation.

A capacitance-voltage measurement also provides the doping density profile of one-sided p-n diodes. For a p^{+}/sup>-

n diode, one obtains the doping density from:

$$N_d = -\frac{2}{q\varepsilon_s}\frac{1}{\frac{d(1/C_j^2)}{dV_a}}, \text{if } N_a >> N_d$$

while the depth equals the depletion layer width, obtained from $x_d = \varepsilon_s A/C_j$. Both the doping density and the corresponding depth can be obtained at each voltage, yielding a doping density profile.

As an example, we consider the measured capacitance-voltage data obtained on a 6H-SiC p-n diode. The diode consists of a highly doped *p*-type region on a lightly doped *n*-type region on top of a highly doped *n*-type substrate. The measured capacitance as well as $1/C^2$is plotted as a function of the applied voltage. The dotted line forms a reasonable fit at voltages close to zero from which one can conclude that the doping density is almost constant close to the p-n interface. The capacitance becomes almost constant at large negative voltages, which corresponds according to equation to a high doping density.

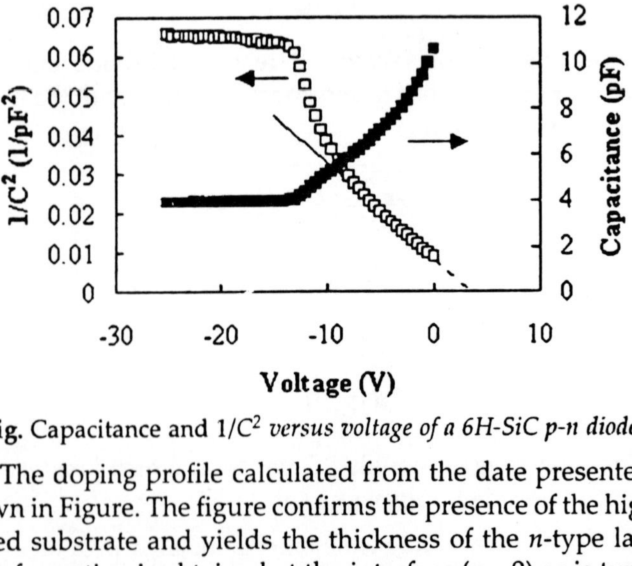

Fig. Capacitance and $1/C^2$ *versus voltage of a 6H-SiC p-n diode.*

The doping profile calculated from the date presented is shown in Figure. The figure confirms the presence of the highly doped substrate and yields the thickness of the *n*-type layer. No information is obtained at the interface ($x = 0$) as is typical for doping profiles obtained from C-V measurements. This is because the capacitance measurement is limited to small forward bias voltages since the forward bias current and the diffusion capacitance affect the accuracy of the capacitance measurement.

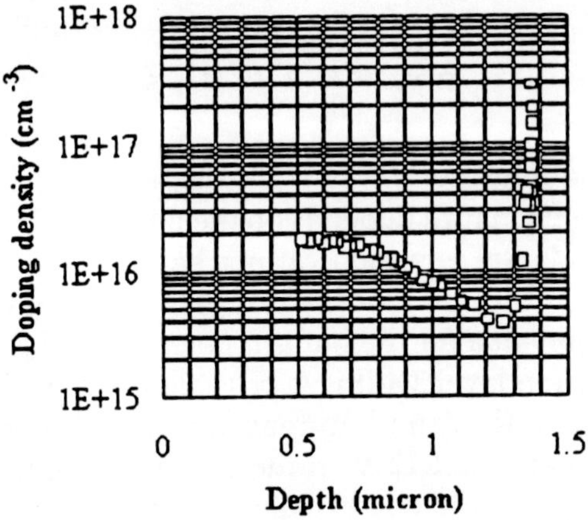

Fig. Doping profile corresponding to the measured data, shown in Figure.

THE LINEARLY GRADED P-N JUNCTION

A linearly graded junction has a doping profile, which depends linearly on the distance from the interface.

$$N_d - N_a = ax$$

To analyse such junction we again use the full depletion approximation, namely we assume a depletion region with width x_n in the n-type region and x_p in the p-type region. Because of the symmetry, we can immediately conclude that both depletion regions must be the same. The potential across the junction is obtained by integrating the charge density between $x = -x_p$ and $x = x_n = x_p$ twice resulting in:

$$\frac{qax^3 p}{3\varepsilon_s} + \frac{qax_n^3}{3\varepsilon_s} = \phi_i - V_a$$

Where the built-in potential is linked to the doping density at the edge of the depletion region such that:

$$x_n = x_p = 3\sqrt{\frac{3\varepsilon_s(\phi_i - V_a)}{2qa}}$$

The depletion layer with is then obtained by solving for the following equation:

$$x_n = x_p = 3\sqrt{\frac{3\varepsilon_s(\phi_i - V_a)}{2qa}}$$

Since the depletion layer width depends on the built-in potential, which in turn depends on the depletion layer width, this transcendental equation cannot be solved analytically. Instead it is solved numerically through iteration. One starts with an initial value for the built-in potential and then solves for the depletion layer width. A possible initial value for the built-in potential is the bandgap energy divided by the electronic charge, or 1.12 V in the case of silicon. From the depletion layer width, one calculates a more accurate value for the built-in potential and repeats the calculation of the depletion layer width. As one repeats this process, one finds that the values for the built-in potential and depletion layer width converge.

The capacitance of a linearly graded junction is calculated like before as:

$$C(V_a) \overset{\Delta}{=} \left|\frac{dQ(V_a)}{dV_a}\right|$$

Where the charge per unit area must be recalculated for the linear junction, namely:

$$|Q(V_a)| = \frac{qax_n^2}{2}$$

The capacitance then becomes:

$$C(V_a) = qax_n\left|\frac{dx_n}{dV_a}\right| = \frac{\varepsilon_s}{2x_n}$$

The capacitance of a linearly graded junction can also be expressed as a function of the zero-bias capacitance or:

$$C(V_a) = \frac{C_{j0}}{\sqrt[3]{1 - \frac{V_a}{\phi_i}}}$$

Where C_{j0} is the capacitance at zero bias, which is given by:

$$C_{j0} = \sqrt[3]{\frac{qa\varepsilon_s^2}{12\phi_i}}$$

THE ABRUPT P-I-N JUNCTION

A p-i-n junction is similar to a p-n junction, but contains in addition an intrinsic or un-intentionally doped region with thickness, d, between the n-type and p-type layer. Such structure is typically used if one wants to increase the width of the depletion region, for instance to increase the optical absorption in the depletion region. Photodiodes and solar cells are therefore likely to be p-i-n junctions.

The analysis is also similar to that of a p-n diode, although the potential across the undoped region, ϕ_u, must be included in the analysis. Equation then becomes:

$$\phi_n + \phi_p + \phi_u = \phi_i - V_a$$

$$\phi_n = \frac{qN_d x_n^2}{2\varepsilon_s}, \phi_p = \frac{qN_a x_p^2}{2\varepsilon_s} \text{ and } \phi_u = \frac{qN_a x_p d}{\varepsilon_s}$$

while the charge in the n-type region still equals that in the p-type region, so that still holds:

$$qN_a x_p = qN_d x_n$$

Equation can be solved for x_n yielding:

$$x_n = \frac{\sqrt{d^2 + \frac{2\varepsilon_s}{q}\frac{(N_a + N_d)}{N_a N_d}(\phi_i - V_a)} - d}{1 + \frac{N_d}{N_a}}$$

From x_n and x_p, all other parameters of the p-i-n junction can be obtained. The total depletion layer width, x_d, is obtained from:

$$x_d = x_n + x_p + d = \sqrt{d^2 + \frac{2\varepsilon_s}{q}\frac{(N_a + N_d)}{N_a N_d}(\phi_i - V_a)}$$

The potential throughout the structure is given by:

$$\phi(x) = -\frac{qN_d}{2\varepsilon_s}(x + x_n)^2 \qquad -x_n < x < 0$$

$$\phi(x) = -\phi_n - \frac{qN_d x_n}{\varepsilon_s}x \qquad 0 < x < d$$

$$\phi(x) = -(\phi_i - V_a) + \frac{qN_a}{2\varepsilon_s}(x - d - x_p)^2 \qquad d < x < d + x_p$$

where the potential at $x = -x_n$ was assumed to be zero.

Capacitance of the p-i-n junction

The capacitance of a p-i-n diode equals the series connection of the capacitances of each region, simply by adding both depletion layer widths and the width of the undoped region:

$$C_j = \frac{\varepsilon_s}{x_n + x_p + d} = \frac{\varepsilon_s}{x_d}$$

SOLUTION TO POISSON'S EQUATION FOR AN ABRUPT P-N JUNCTION

Applying Gauss's law one finds that the total charge in the n-type depletion region equals minus the charge in the p-type depletion region:

$$Q_n = \varepsilon_s \,|\, \mathcal{E}(x=0) \,| = -Q_p$$

Poisson's equation can be solved separately in the *n*-type and *p*-type region yielding an expression for $\mathcal{E}(x = 0)$ which is almost identical to equation:

$$|\mathcal{E}(x=0)| = \frac{V_t}{L_{D,n}}\sqrt{2(\exp\frac{\phi_n}{V_t} - \frac{\phi_n}{V_t} - 1)} = \frac{V_t}{L_{D,n}}\sqrt{2(\exp\frac{\phi_p}{V_t} - \frac{\phi_p}{V_t} - 1)}$$

where ϕ_n and ϕ_p are assumed negative if the semiconductor is depleted. Their relation to the applied voltage is given by:

$$\phi_n + \phi_p = V_a - \phi_i$$

One obtains ϕ_n and ϕ_p as a function of the applied voltage by solving the transcendental equations.

For the special case of a symmetric doping profile, or $N_d = N_a$, these equations can easily be solved yielding:

$$\phi_n = \phi_p = \frac{V_a - \phi_i}{2}$$

The depletion layer widths also equal each other and are given by:

$$x_n = x_p = \left|\frac{Q_n}{qN_d}\right| = \left|\frac{Q_p}{qN_a}\right| = \left|\frac{\varepsilon_s \mathcal{E}(x=0)}{qN_d}\right|$$

Using the above expression for the electric field at the origin, we find:

$$x_n = x_p = L_D\sqrt{2\exp\frac{V_a - \phi_i}{2V_t} + \frac{\phi_i - V_a - V_t}{V_t}}$$

where $L_{D,n} = L_{D,p} = \sqrt{\frac{\varepsilon_s kT}{q^2 Nd}}$ is the extrinsic Debye length.

The relative error of the depletion layer width as obtained using the full depletion approximation equals:

$$\frac{\Delta x_n}{x_n}=\frac{\Delta x_p}{x_p}=\frac{1-\sqrt{1-\dfrac{V_t}{\phi_i-V_a}+\dfrac{V_t}{2(\phi_i-V_a)}\exp\left(\dfrac{V_a-\phi_i}{2V_t}\right)}}{\sqrt{1-\dfrac{V_t}{\phi_i-V_a}+\dfrac{V_t}{2(\phi_i-V_a)}\exp\left(\dfrac{V_a-\phi_i}{2V_t}\right)}}$$

So that for $\frac{\phi_i - V_a}{V_t}$ = 1, 2, 5, 10, 20 and 40, one finds the relative error to be 45, 23, 10, 5.1, 2.5 and 1.26 per cent.

THE HETERO P-N JUNCTION

Heterojunction p-n diodes can be found in a wide range of heterojunction devices including laser diodes, high electron mobility transistors (HEMTs) and heterojunction bipolar transistors (HBTs). Such devices take advantage of the choice of different materials, and the corresponding material properties, for each layer of the heterostructure. We present in this section the electro-static analysis of heterojunction p-n diodes.

The heterojunction p-n diode is in principle very similar to a homojunction. The main problem that needs to be tackled is the effect of the bandgap discontinuities and the different material parameters, which make the actual calculations more complex even though the p-n diode concepts need almost no changing. An excellent detailed treatment can be found in Wolfe et al.

Band diagram of a Heterojunction p-n diode under Flatband Conditions

The flatband energy band diagram of a heterojunction p-n diode is shown in the figure below. As a convention we will assume ΔE_c to be positive if $E_{c,n} > E_{c,p}$ and ΔE_v to be positive if $E_{v,n} < E_{v,p}$.

Calculation of the Contact Potential (built-in voltage)

The built-in potential is defined as the difference between the Fermi levels in both the n-type and the p-type semiconductor. From the energy diagram we find:

$$q\phi_i = E_{F,n} - E_{F,p}$$

which can be expressed as a function of the electron concentrations and the effective densities of states in the conduction band:

$$q\phi_i = \Delta E_c + kTln\frac{n_{n0}N_{c,p}}{n_{p0}N_{c,n}}$$

The built-in voltage can also be related to the hole concentrations and the effective density of states of the valence band:

$$q\phi_i = -\Delta E_v + kT\ln\frac{p_{p0}N_{v,n}}{p_{n0}N_{v,p}}$$

Combining both expressions yields the built-in voltage independent of the free carrier concentrations:

$$q\phi_i = \frac{\Delta E_c - \Delta E_v}{2} + kT\ln\frac{N_d N_a}{n_{i,n}n_{i,p}} + \frac{kT}{2}\ln\frac{N_{v,n}N_{c,p}}{N_{c,n}N_{v,p}}$$

where $n_{i,n}$ and $n_{i,p}$ are the intrinsic carrier concentrations of the *n*-type and *p*-type region, respectively. ΔE_c and ΔE_v are positive quantities if the bandgap of the *n*-type region is smaller than that of the *p*-type region and the sum of both equals the bandgap difference. The band alignment must also be as shown in Figure.The above expression reduces to that of the built-in junction of a homojunction if the material parameters in the *n*-type region equal those in the *p*-type region. If the effective densities of states are the same, the expression for the heterojunction reduces to:

$$q\phi_i = \frac{\Delta E_c - \Delta E_v}{2} + kT\ln\frac{N_d N_a}{n_{i,n}n_{i,p}}$$

Abrupt p-n Heterojunction

For the calculation of the charge, field and potential distribution in an abrupt p-n junction we follow the same approach as for the homojunction. First of all we use the full depletion approximation and solve Poisson's equation.

$$\phi_n + \phi_p = \phi_i - V_a$$

$$\phi_n = \frac{qN_d x_n^2}{2\varepsilon_{s,n}} \text{ and } \phi_p = \frac{qN_a x_n^2}{2\varepsilon_{s,p}}$$

$$qN_a x_p = qN_d x_n$$

The main differences are the different expression for the built-in voltage and the discontinuities in the field distribution (because of the different dielectric constants of the two regions) and in the energy band diagram. However the expressions for x_n and x_p for a homojunction can still be used if one replaces N_a by $N_a\ \varepsilon_{s,p}/\varepsilon_s$, N_d by $N_d\ \varepsilon_{s,n}/\varepsilon_s$, x_p by $x_p\ \varepsilon_s/\varepsilon_{s,p}$, and x_n by $x_n\ \varepsilon_s/\varepsilon_{s,n}$. Adding x_n and x_p yields the total depletion layer width x_d:

$$x_d = x_n + x_p = \sqrt{\frac{2\varepsilon_{s,n}\varepsilon_{s,p}}{q}\frac{(N_a + N_d)^2(\phi_i - V_a)}{N_a N_d(N_a\varepsilon_{s,p} + N_d\varepsilon_{s,n})}}$$

The capacitance per unit area can be obtained from the series connection of the capacitance of each layer:

$$C_j = \frac{1}{\dfrac{x_n}{\varepsilon_{s,n}} + \dfrac{x_p}{\varepsilon_{s,p}}} = \sqrt{\frac{q\varepsilon_{s,n}\varepsilon_{s,p}}{2}\frac{N_a N_d}{(N_a\varepsilon_{s,p} + N_d\varepsilon_{s,n})(\phi_i - V_a)}}$$

Abrupt P-i-N heterojunction

For a P-i-N heterojunction the above expressions take the following modified form:

$$\phi_n + \phi_p + \phi_u = \phi_i - V_a$$

$$\phi_n = \frac{qN_d x_n^2}{2\varepsilon_{s,n}},\ \phi_p = \frac{qN_a x_p^2}{2\varepsilon_{s,p}} \text{ and } \phi_u = \frac{qN_a x_p d}{2\varepsilon_{s,u}}$$

$$qN_a x_p = qN_d x_n$$

Where ϕ_u is the potential across the middle undoped region of the diode, having a thickness *d*. The depletion layer width and the capacitance are given by:

$$x_d = x_n + x_p + d$$

$$C_j = \frac{1}{\dfrac{x_n}{\varepsilon_{s,n}} + \dfrac{x_p}{\varepsilon_{s,p}} + \dfrac{d}{\varepsilon_s}}$$

Equations can be solved for x_n, yielding:

$$x_n = \frac{\sqrt{\left(\dfrac{d\,\varepsilon_{s,n}}{\varepsilon_{s,u}}\right)^2 + \dfrac{2\varepsilon_{s,n}(\phi_i - V_a)}{qN_d}\left(1 + \dfrac{N_d}{N_a}\right)} - \dfrac{d\,\varepsilon_{s,n}}{\varepsilon_{s,u}}}{\left(1 + \dfrac{N_d}{N_a}\right)}$$

A solution for x_p can be obtained from by replacing N_d by N_a, N_a by N_d, $\varepsilon_{s,n}$ by $\varepsilon_{s,p}$, and $\varepsilon_{s,p}$ by $\varepsilon_{s,n}$. Once x_n and x_p are determined all other parameters of the P-i-N junction can be obtained. The potential throughout the structure is given by:

$$\phi(x) = -\frac{qN_d}{2\varepsilon_{s,n}}(x + x_n)^2 \text{ for } -x_n < x < 0$$

$$\phi(x) = -\phi_n - \frac{qN_d x_n}{2\varepsilon_{s,n}} x \text{ for } 0 < x < d$$

$$\phi(x) = -(\phi_i - V_a) + \frac{qN_a}{2\varepsilon_{s,p}}(x - d - x_p)^2 \text{ for } d < x < d + x_p$$

where the potential at $x = -x_n$ was assumed to be zero.

An example of the charge distribution, electric field, potentials and energy band diagram throughout the P-i-N heterostructure is presented in Figure:

The above derivation ignores the fact that - because of the energy band discontinuities - the carrier densities in the intrinsic region could be substantially larger than in the depletion regions in the *n*-type and *p*-type semiconductor.

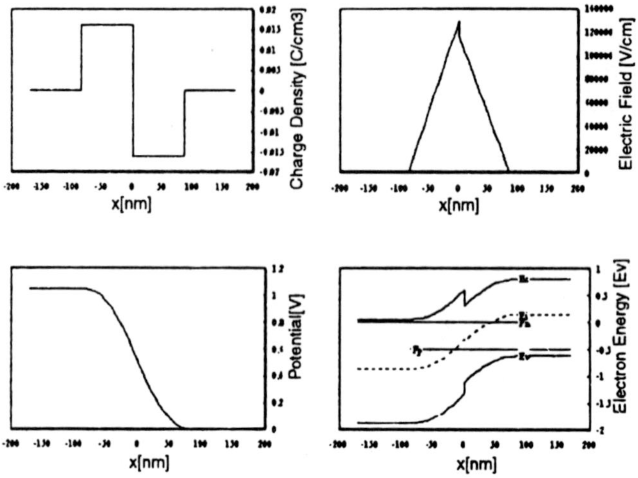

Fig. Charge distribution, electric field, potential and energy band diagram of an AlGaAs/GaAs p-n heterojunction with $V_a = 0.5$ V, $x = 0.4$ on the left and $x = 0$ on the right. $N_d = N_a = 10^{17}\text{cm}^{-3}$

Large amounts of free carriers imply that the full depletion approximation is not valid and that the derivation has to be repeated while including a possible charge in the intrinsic region.

A P-M-N Junction with Interface Charges

Real P-i-N junctions often differ from their ideal model, the intrinsic region could be lightly doped, while a fixed interface charge could be present between the individual layers. We now consider the middle layer to have a doping concentration $N_m = N_{dm} - N_{am}$ and a dielectric constant $\varepsilon_{s,m}$. A charge Q_1 is assumed between the N and M layer, and a charge Q_2 between the M and P layer. Equation then take the following form:

$$\phi_n + \phi_p + \phi_m = \phi_i - V_a$$

$$\phi_n = \frac{qN_d x_n^2}{2\varepsilon_{s,n}},\ \phi_p = \frac{qN_a x_p^2}{2\varepsilon_{s,p}} \text{ and } \phi_m = (qN_a x_p + Q_1)\frac{d}{\varepsilon_{s,m}} + \frac{qN_m d^2}{2\varepsilon_{s,m}}$$

$$qN_d x_n + Q_1 + Q_2 + qN_m d = qN_a x_p$$

These equations can be solved for x_n and x_p yielding a general solution for this structure. Again it should be noted that this solution is only valid if the middle region is indeed fully depleted.

Solving the above equation allows to draw the charge density, the electric field distribution, the potential and the energy band diagram. An example is provided in Figure.

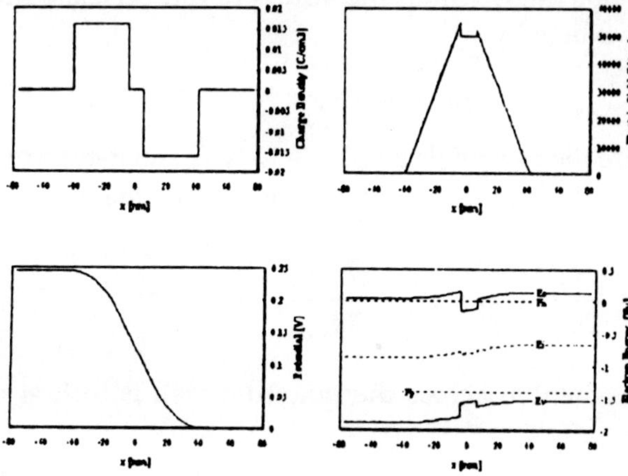

Fig. Charge distribution, electric field, potential and energy band diagram of an AlGaAs/GaAs p-i-n heterojunction with $V_a = 1.4$ V, $x = 0.4$ on the left, $x = 0$ in the middle and $x = 0.2$ on the right. $d = 10$ nm and $N_d = N_a = 10^{17}\text{cm}^{-3}$

Quantum Well in a p-n Junction

Next, we consider a p-n junction with a quantum well located between the *n* and *p* region as shown in Figure.

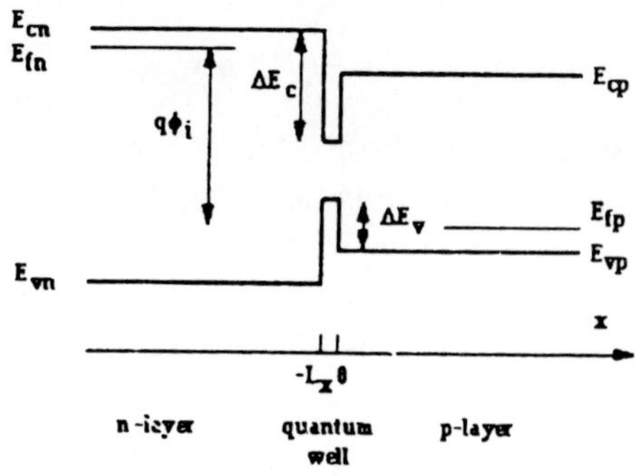

Fig. Flat-band energy band diagram of a p-n heterojunction with a quantum well at the interface.

Under forward bias, charge can accumulate within the quantum well. In this section, we will outline the procedure to solve this structure. The actual solution can only be obtained by solving a transcendental equation. Approximations will be made to obtain useful analytic expressions.

The potentials within the structure can be related to the applied voltage by:

$$\phi_n + \phi_{qw} + \phi_p = \phi_i - V_a$$

where the potentials across the *p*-type and *n*–type regions are obtained using the full depletion approximation:

$$\phi_n = \frac{qN_d x_n^2}{2\varepsilon_{s,n}}, \text{ and } \phi_p = \frac{qN_a x_p^2}{2\varepsilon_{s,p}},$$

The potential across the quantum well is to first order given by:

$$q_{qw} = \frac{qN_d x_n L_x}{2\varepsilon_{s,n}} + \frac{q(P-N)L_x}{2\varepsilon_{s,p}}$$

where *P* and *N* are the hole and electron density per unit area in the quantum well. This equation assumes that the charge in the quantum well $Q = q\ (P - N)$ is located in the middle of the well. Applying Gauss's law yields the following balance between the charges:

$$qN_d x_n - qN = -qP + qN_a x_p$$

where the electron and hole densities can be expressed as a function of the effective densities of states in the quantum well:

$$N = N_{c,qw} \sum_{n=1}^{\infty} \ln[1 + \exp \frac{\Delta E_{n,e}}{kT}]$$

$$P = P_{v,qw} \sum_{n=1}^{\infty} \ln[1 + \exp \frac{\Delta E_{n,h}}{kT}]$$

with $\Delta E_{n,e}$ and $\Delta E_{n,h}$ given by:

$$\Delta E_{n,e} = \Delta E_c - q\phi_n - kT\ln\frac{N_{c,n}}{N_d} - E_{n,e}$$

$$\Delta E_{n,h} = \Delta E_v - q\phi_p - kT\ln\frac{N_{v,p}}{N_a} - E_{n,h}$$

where $E_{n,e}$ and $E_{n,h}$ are the nth energies of the electrons respectively holes relative to the conduction respectively valence band edge. These nine equations can be used to solve for the nine unknowns by applying numerical methods. A quick solution can be obtained for a symmetric diode, for which all the parameters (including material parameters) of the n and p region are the same. For this diode N equals P because of the symmetry. Also x_n equals x_p and ϕ_n equals ϕ_p. Assuming that only one energy level namely the $n = 1$ level is populated in the quantum well one finds:

$$N = P = N_{c,qw}\ln[1 + \exp\frac{qV_a - 2E_{1,e} - E_g}{2kT}]$$

where E_g is the bandgap of the quantum well material.

Numeric simulations of the general case reveal that, especially under large forward bias conditions, the electron and hole density in the quantum well are the same to within a few percent. An example is presented in Figure.

From the numeric simulation of a GaAs n-qw-p structure we find that typically only one electron level is filled with electrons, while several hole levels are filled with holes or

$$N = N_1 \cong P = P_1 + P_2 + P_3 + \ldots.$$

If all the quantized hole levels are more than $3kT$ below the hole quasi-Fermi level one can rewrite the hole density as:

$$P = P_1 \sum_n \exp\frac{E_{1,h} - E_{n,h}}{kT}$$

Since the 2-D densities of states are identical for each

quantized level. The applied voltage is given by:

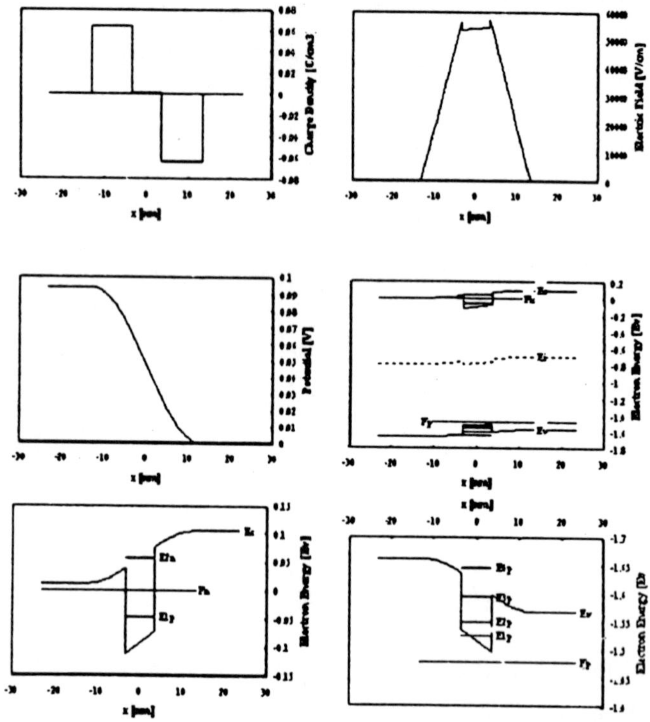

Fig. Energy band diagram of a GaAs/AlGaAs p-n junction with a quantum well in between. The aluminum concentration is 40 per cent for both the *p* and *n* region, and zero in the well. The doping concentrations N_a and N_d are 4 x 10^{17} cm^{-3} and V_a = 1.4 V.

$$V_a = \frac{E_{g,qwl}}{q} + V_t \ln(e^{N/N_{c,qw}} - 1)(e^{N/N^*_{v,qw}} - 1)$$

with

$$N^*_{v,qw} = N_{v,qw} \sum_n \exp \frac{E_{1,h} - E_{n,h}}{kT}$$

or

$$N^*_{v,qw} = N_{v,qw}(1 + \exp \frac{-3E_{1,h}}{kT} + \exp \frac{-8E_{1,h}}{kT} + \exp \frac{-15E_{1,h}}{kT} + \ldots)$$

THE P-N DIODE CURRENT

GENERAL DISCUSSION

The current in a p-n diode is due to carrier recombination or generation somewhere within the p-n diode structure. Under forward bias, the diode current is due to recombination. This recombination can occur within the quasi-neutral regions, within the depletion region or at the metal-semiconductor Ohmic contacts. Under reverse bias, the current is due to generation. Carrier generation due to light will further increase the current under forward as well as reverse bias.

We first derive the ideal diode current. We will also distinguish between the "long" diode and "short" diode case. The "long" diode expression applies to p-n diodes in which recombination/generation occurs in the quasi-neutral region only. This is the case if the quasi-neutral region is much larger than the carrier diffusion length. The "short" diode expression applies to p-n diodes in which recombination/generation occurs at the contacts only. In a short diode, the quasi-neutral region is much smaller than the diffusion length. In addition to the "long" and "short" diode expressions, we also present the general result for p-n diodes with arbitrary widths.

Next, we derive expressions for the recombination/ generation in the depletion region. Here we have to distinguish between the different recombination mechanisms - band-to-band recombination and Shockley-Hall-Read recombination - as they lead to different current-voltage characteristics.

THE IDEAL DIODE CURRENT

General Discussion and Overview

When calculating the current in a p-n diode one needs to know the carrier density and the electric field throughout the p-n diode which can then be used to obtain the drift and diffusion current. Unfortunately, this requires the knowledge of the quasi-Fermi energies, which is only known if the currents are known. The straightforward approach is to simply solve the drift-diffusion simultaneously. This approach however does not yield an analytic solution.

To avoid this problem we will assume that the electron and hole quasi-Fermi energies in the depletion region equal those in the adjacent n-type and p-type quasi-neutral regions. We will derive an expression for "long" and "short" diodes as well as a general expression, which is to be used if the quasi-neutral region is comparable in size to the diffusion length.

Assumptions and Boundary Conditions

The electric field and potential are obtained by using the full depletion approximation. Assuming that the quasi-Fermi energies are constant throughout the depletion region, one obtains the minority carrier densities at the edges of the depletion region, yielding:

$$p_n(x = x_n) = p_{n0}\, e^{V_a / V_t}$$

and

$$n_p(x = -x_p) = n_{p0}\, e^{V_a / V_t}$$

These equations can be verified to yield the thermal-equilibrium carrier density for zero applied voltage. In addition, an increase of the applied voltage will increase the separation between the two quasi-Fermi energies by the applied voltage multiplied with the electronic charge.

The carrier density at the metal contacts is assumed to equal the thermal-equilibrium carrier density. This assumption implies that excess carriers immediately recombine when reaching either of the two metal-semiconductor contacts. As recombination is typically higher at a semiconductor surface and is further enhanced by the presence of the metal, this is found to be a reasonable assumption. This results in the following set of boundary conditions:

$$p_n\left(w = w_n\right) = p_{n0}$$

and

$$n_p\left(x = -w_p\right) = n_{p0}$$

General Current Expression

The general expression for the ideal diode current is obtained by applying the boundary conditions to the general solution of the diffusion equation for each of the quasi-neutral regions, as described by equations:

$$p_n(x \geq x_n) = p_{n0} + Ae^{-(x-x_n)/L_p} + Be^{(x-x_n)/L_p}$$

$$n_p(x \leq -x_p) = n_{p0} + Ce^{-(x+x_p)/L_p} + De^{(x+x_p)/L_p}$$

The boundary conditions at the edge of the depletion regions are described by equations.

Before applying the boundary conditions, it is convenient to rewrite the general solution in terms of hyperbolic functions:

$$p_n(x \geq x_n) = p_{n0} + A^* \cosh \frac{x - x_n}{L_p} + B^* \sinh \frac{x - x_n}{L_p}$$

$$n_p(x \leq -x_p) = n_{p0} + C^* \cosh \frac{x + x_p}{L_n} + D^* \sinh \frac{x + x_p}{L_n}$$

where A^*, B^*, C^* and D^* are constants whose value remains to be determined. Applying the boundary conditions then yields:

$$p_n(x \geq x_n) = p_{n0} + p_{n0}(e^{V_a/V_t} - 1)[\cosh \frac{x - x_n}{L_p} - \coth \frac{w_n'}{L_p} \sinh \frac{x - x_n}{L_p}]$$

$$n_p(x \leq -x_p) = n_{p0} + n_{p0}(e^{V_a/V_t} - 1)[\cosh \frac{x + x_p}{L_n} + \coth \frac{w_p'}{L_n} \sinh \frac{x + x_p}{L_n}]$$

Where the quasi-neutral region widths, w_n' and w_p', are defined as:

$$w_n' = w_n - x_n$$

and

$$w_p' = w_p - x_p$$

The current density in each region is obtained by calculating the diffusion current density using equations:

$$J_p(x \geq x_n) = -qD_p \frac{dp}{dx}$$

$$= -\frac{qD_p p_{n0}}{L_p}(e^{V_a/V_t} - 1)[\sinh \frac{x - x_n}{L_p} - \coth \frac{w_n'}{L_p} \cosh \frac{x - x_n}{L_p}]$$

$$J_n(x \leq -x_p) = qD_n \frac{dn}{dx}$$

$$= \frac{qD_n n_{p0}}{L_n}(e^{V_a/V_t} - 1)[\sinh \frac{x + x_p}{L_n} + \coth \frac{w_p'}{L_n} \cosh \frac{x + x_p}{L_n}]$$

The total current must be constant throughout the structure since a steady state case is assumed. No charge can accumulate or disappear somewhere in the structure so that the charge flow must be constant throughout the diode. The total current then equals the sum of the maximum electron current in the p-type region, the maximum hole current in the n-type regions and the current due to recombination within the depletion region. The maximum currents in the quasi-neutral regions occur at either side of the depletion region and can therefore be calculated from equations. Since we do not know the current due to recombination in the depletion region we will simply assume that it can be ignored. Later, we will more closely examine this assumption. The total current is then given by:

$$I = A[J_n(x = -x_p) + J_p(x = x_n) + J_r] \cong I_s(e^{V_a/V_t} - 1)$$

where I_s can be written in the following form:

$$I_s = qA[\frac{D_n n_{p0}}{L_n} \coth(\frac{w_p'}{L_n}) + \frac{D_p p_{n0}}{L_p} \coth(\frac{w_n'}{L_p})]$$

The p-n Diode with a "long" Quasi-neutral Region

A diode with a "long" quasi-neutral region has a quasi-neutral region, which is much larger than the minority-carrier diffusion length in that region, or $w_n' > L_p$ and $w_p' > L_n$. The general solution can be simplified under those conditions using:

$$\coth x = \frac{1}{\tanh x} \cong \frac{1}{x}, \text{for } x \ll 1$$

yielding the following carrier densities, current densities and saturation currents:

$$p_n(x \geq x_n) = p_{n0} + p_{n0}(e^{V_a/V_t} - 1)\exp\frac{-(x - x_n)}{L_p}$$

$$n_p(x \leq -x_p) = n_{p0} + n_{p0}(e^{V_a/V_t} - 1)\exp\frac{x + x_p}{L_n}$$

$$J_p(x \geq x_n) = \frac{qD_p p_{n0}}{L_p}(e^{V_a/V_t} - 1)\exp\frac{-(x - x_n)}{L_p}$$

$$J_n(x \leq -x_p) = \frac{qD_n n_{p0}}{L_n}(e^{V_a/V_t} - 1)\exp\frac{x + x_p}{L_n}$$

$$I_s = qA[\frac{D_n n_{p0}}{L_n} + \frac{D_p p_{n0}}{L_p}] = qA[\frac{n_{p0} L_n}{\tau_n} + \frac{p_{n0} L_p}{\tau_p}]$$

We now come back to our assumption that the current due to recombination in the depletion region can be simply ignored. Given that there is recombination in the quasi-neutral region, it would be unreasonable to suggest that the recombination rate would simply drop to zero in the depletion region. Instead, we assume that the recombination rate is constant in the depletion region. To further simplify the analysis we will consider a p^+-n junction so that we only need to consider the recombination in the n-type region. The current due to recombination in the depletion region is then given by:

$$I_r \geq qA\frac{p_{n0} x_n}{\tau_p}(e^{V_a/V_t} - 1)$$

so that I_r can be ignored if:

$$I_r \ll I, \text{for } x_n \ll L_p$$

A necessary, but not sufficient requirement is therefore that the depletion region width is much smaller than the diffusion length for the ideal diode assumption to be valid. Silicon and germanium p-n diodes usually satisfy this requirement, while gallium arsenide p-n diodes rarely do because of the short carrier lifetime and diffusion length.

As an example we now consider a silicon p-n diode with $N_a = 1.5 \times 10^{14}$ cm^{-3} and $N_d = 10^{14}$ cm^{-3}. The minority carrier lifetime was chosen to be very short, namely 400 ps, so that most features of interest can easily be observed. We start by examining the electron and hole density throughout the p-n diode, shown in Figure:

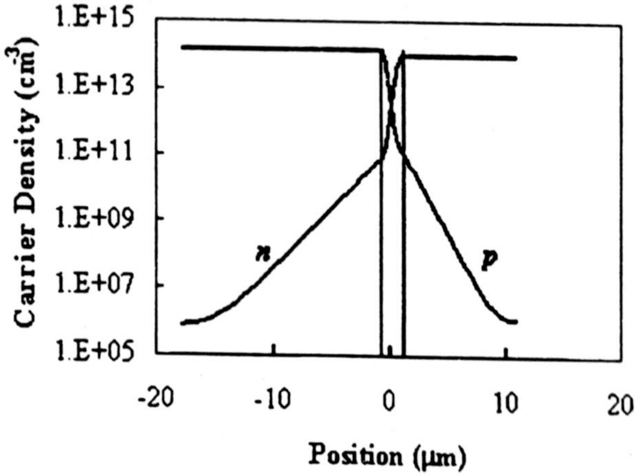

Fig. Electron and hole density throughout a forward biased p-n diode.

The majority carrier densities in the quasi-neutral region simply equal the doping density. The minority carrier densities in the quasi-neutral regions are obtained from equations. The electron and hole densities in the depletion region are calculating using the assumption that the electron/hole quasi-Fermi energy in the depletion region equals the electron/hole quasi-Fermi energy in the quasi-neutral *n*-type/*p*-type region. The corresponding band diagram is shown in Figure:

The quasi-Fermi energies were obtained by combining equations. Note that the quasi-Fermi energies vary linearly within the quasi-neutral regions.

Shown in Figure is the electron and hole current density as calculated using equations.

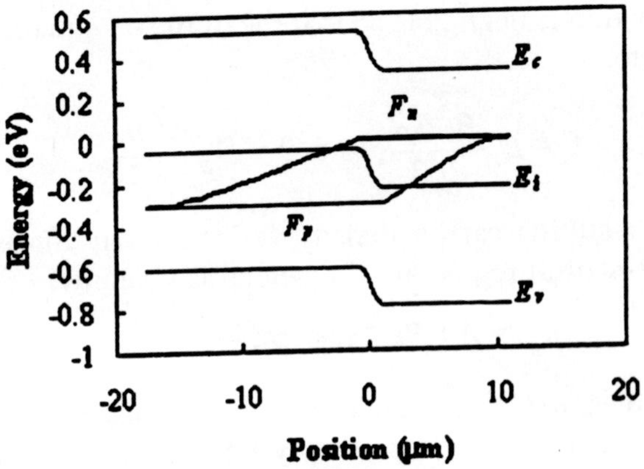

Fig. Energy band diagram of a p-n diode. Shown are the conduction band edge, E_c, and the valence band edge, E_v, the intrinsic energy, E_i, the electron quasi-Fermi energy, F_n, and the hole quasi-Fermi energy, F_p.

The current due to recombination in the depletion region was assumed to be constant.

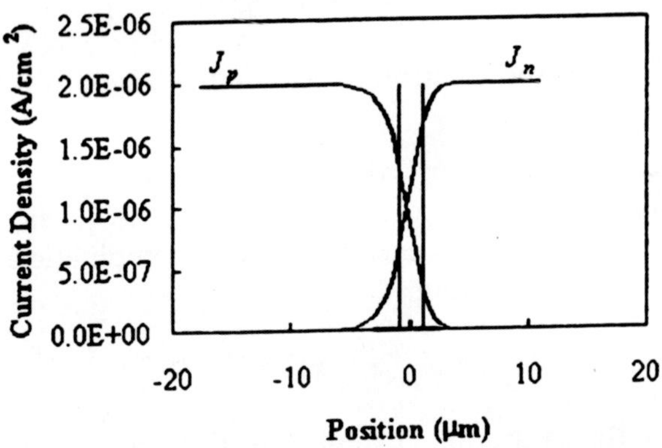

Fig. Electron and hole current density versus position. The vertical lines indicate the edges of the depletion region.

The p-n Diode with a "short" Quasi-neutral Region

A "short" diode is a diode with quasi-neutral regions, which are much shorter than the minority-carrier diffusion lengths. As the quasi-neutral region is much smaller than the diffusion length one finds that recombination in the quasi-neutral region is negligible so that the diffusion equations are reduced to:

$$0 = D_n \frac{d^2 n_p}{dx^2}, \text{ and } 0 = D_p \frac{d^2 p_n}{dx^2}$$

The resulting carrier density varies linearly throughout the quasi-neutral region and in general is given by:

$$n_p = A + Bx\text{, and } p_n = A + Bx$$

where *A*, *B*, *C* and *D* are constants obtained by satisfying the boundary conditions. Applying the same boundary conditions at the edge of the depletion region as above and requiring thermal equilibrium at the contacts yields:

$$p_n = p_{n0} + p_{n0}(e^{V_a/V_t} - 1)\left(1 - \frac{x - x_n}{w_n'}\right)$$

$$n_p = n_{p0} + n_{p0}(e^{V_a/V_t} - 1)\left(1 + \frac{x + x_p}{w_p'}\right)$$

for the hole and electron density in the *n*-type and *p*-type quasi-neutral region.

The current in a "short" diode is again obtained by adding the maximum diffusion currents in each of the quasi-neutral regions and ignoring the current due to recombination in the depletion region, yielding:

$$I = A[J_n(x = -x_p) + J_p(x = x_n) + J_r] \cong I_s(e^{V_a/V_t} - 1)$$

where the saturation current, I_s is given by:

$$I_s = qA[\frac{D_n n_{p0}}{w_p'} + \frac{D_p p_{n0}}{w_n'}]$$

A comparison of the "short" diode expression with the "long" diode expression reveals that they are the same except for the use of either the diffusion length or the quasi-neutral region width in the denominator, whichever is smaller.

Now that we have two approximate expressions, it is of interest to know when to use one or the other. To this end, we now consider a one-sided n^+-p diode. The *p*-type semiconductor has a width, w_p, and we normalize the excess electron density relative to its value at the edge of the depletion region ($x = 0$). The Ohmic contact to the *p*-type region is ideal so that the excess density is zero at $x = w_p'$. The normalized excess carrier density is shown in Figure for different values of the diffusion length.

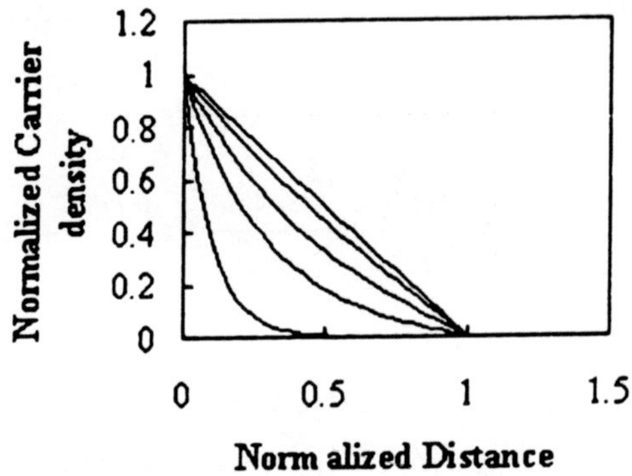

Fig. Excess electron density versus position as obtained by solving the diffusion equation with $\delta n(x = 0) = 1$ and $\delta n(x/w_p' = 1) = 0$. The ratio of the diffusion length to the width of the quasi-neutral region is varied from 0.1 (Bottom curve), 0.3, 0.5, 1 and ∞ (top curve)

The figure illustrates how the excess electron density changes as the diffusion length is varied relative to the width of the quasi-neutral region. For the case where the diffusion length is much smaller than the width ($L_n << w_p'$), the electron density decays exponentially and the "long" diode expression can be used. If the diffusion length is much longer than the width ($L_n >> w_p'$), the electron density reduces linearly with position and the "short" diode expression can be used. If the diffusion length is comparable to the width of the quasi-neutral region width one must use the general expression. A numeric analysis reveals that the error is less than 10 per cent when using the short diode expression with $L_n > 2\, w_p'$ and when using the long diode expression with $L_n < w_p'/2$. Note that the best approximation is not necessarily the same in each region of the same p-n diode.

Example

An abrupt silicon p-n junction ($N_a = 10^{16}$ cm^{-3} and $N_d = 4 \times 10^{16}$ cm^{-3}) is biased with $V_a = 0.6$ V. Calculate the ideal diode current assuming that the n-type region is much smaller than the diffusion length with $w_n' = 1$ μm and assuming a "long" p-type region. Use $\mu_n = 1000$ cm^2/V-s and $\mu_p = 300$ cm^2/V-s. The minority carrier lifetime is 10 μs and the diode area is 100 μm by 100 μm.

SolutionThe current is calculated from:

$$I = qA[\frac{D_n n_{p0}}{L_n} + \frac{D_p p_{n0}}{w_n'}](e^{V_a/V_t} - 1)$$

with

- $D_n = m_n V_t = 1000 \times 0.0258 = 25.8$ cm^2/V-s
- $D_p = m_p V_t = 300 \times 0.0258 = 7.75$ cm^2/V-s
- $n_{p0} = n_i^2/N_a = 10^{20}/10^{16} = 10^4$ cm^{-3}
- $p_{n0} = n_i^2/Nd = 10^{20}/4 \times 10^{16} = 2.5 \times 10^3$ cm^{-3}

$$L_n = \sqrt{D_n \tau_n} = \sqrt{25.8 \times 10^{-5}} = 161\,\mu m$$

yielding $I = 40.7$ mA

Note that the hole diffusion current occurs in the "short" *n*-type region and therefore depends on the quasi-neutral width in that region. The electron diffusion current occurs in the "long" *p*-type region and therefore depends on the electron diffusion length in that region.

RECOMBINATION-GENERATION CURRENT

We now calculate the recombination-generation current in the depletion region of a p-n junction. We distinguish between two different possible recombination mechanisms: band-to-band recombination and Shockley-Hall-Read recombination.

Band-to-band Recombination-Generation Current

The recombination/generation current due to band-to-band recombination/generation is obtained by integrating the net recombination rate, $U_{b\text{-}b}$, over the depletion region:

$$J_{b-b} = q \int_{-x_p}^{x_n} U_{b-b} dx$$

where the net recombination rate is given by:

$$U_{b-b} = b(np - n_i^2)$$

The carrier densities can be related to the constant quasi-Fermi energies and the product is independent of position:

$$np = n_i e^{(E_{Fn} - E_i)/kT} n_i e^{(E_i - E_{Fp})/kT} = n_i^2 e^{V_a/V_t}$$

This allows the integral to be solved analytically yielding:

$$J_{b-b} = q \int_{-x_p}^{x_n} n_i^2 (e^{V_a/V_t} - 1) dx = q n_i^2 bw(e^{V_a/V_t} - 1)$$

The current due to band-to-band recombination has therefore the same voltage dependence as the ideal diode current and simply adds an additional term to the expression for the saturation current.

Shockley-Hall-Read Recombination-Generation current

The current due to trap-assisted recombination in the depletion region is also obtained by integrating the trap-assisted recombination rate over the depletion region width:

$$J_{SHR} = q \int_{-x_p}^{x_n} U_{SHR} dx$$

Substituting the expression for the recombination rate yields:

$$J_{SHR} = q \int_{-x_p}^{x_n} \frac{1}{\tau} \frac{n_i^2 (e^{V_a/V_t} - 1)}{n + p + 2n_i \cosh(\frac{E_t - E_i}{kT})} dx$$

where the product of the electron and hole densities was obtained by assuming that the quasi-Fermi energies are constant throughout the depletion region, which leads to:

$$np = n_i e^{(F_n - E_i)/kT} n_i e^{(E_i - F_p)/kT} = n_i^2 e^{V_a/V_t}$$

The maximum recombination rate is obtained when the electron and hole densities are equal and therefore equals the square root of the product yielding:

$$U_{SHR,\max} = \text{MAX}\left(\frac{1}{\tau} \cdot \frac{n_i^2 (e^{V_a/V_t} - 1)}{n + p + 2n_i \cosh(\frac{E_t - E_i}{kT})}\right) \cong \frac{n_i}{2\tau}(e^{V_a/2V_t} - 1)$$

From which an effective width can be defined which, when multiplied with the maximum recombination rate, equals the integral of the recombination rate over the depletion region. This effective width, x', is then defined by:

$$x' = \frac{\int_{-x_p}^{x_n} \frac{1}{\tau} \frac{n_i^2 (e^{V_a/V_t} - 1)}{n + p + 2n_i \cosh(\frac{E_t - E_i}{kT})} dx}{U_{SHR,\max}}$$

and the associated current due to trap-assisted recombination in the depletion region is given by:

$$J_{SHR} = \frac{q n_i x'}{2\tau}(e^{V_a/2V_t} - 1)$$

This does not provide an actual solution since the effective width, x', still must be determined by performing a numeric integration. Nevertheless, the above expression provides a way to obtain an upper estimate by substituting the depletion layer width, x_d, as it is always larger than the effective width.

I-V CHARACTERISTICS OF REAL P-N DIODES

The forward biased I-V characteristics of real p-n diodes are further affected by high injection and the series resistance of the diode. To illustrate these effects while summarizing the current mechanisms discussed previously we consider the *I-V* characteristics of a silicon p^+-n diode with $N_d = 4 \times 10^{14}$ cm^{-3}, $\tau_p = 10$ μs, and $\mu_p = 450$ cm^2/V-s. The *I-V* characteristics are plotted on a semi-logarithmic scale and four different regions can be distinguished as indicated on Figure. First, there is the ideal diode region where the current increases by one order of magnitude as the voltage is increased by 60 mV. This region is referred to as having an ideality factor, *n*, of one. This ideality factor is obtained by fitting a section of the curve to the following expression for the current:

$$J = J_s e^{V_a / \eta V_t}$$

The ideality factor can also be obtained from the slope of the curve on a semi-logarithmic scale using:

$$\eta = \frac{\log(e)}{V_t \text{ slope}} = \frac{\frac{1}{\text{slope}}}{59.6 \text{ mV/decade}}$$

where the slope is in units of V/decade. To the left of the ideal diode region there is the region where the current is dominated by the trap-assisted recombination in the depletion region. This part of the curve has an ideality factor of two. To the right of the ideal diode region, the current becomes limited by high injection effects and by the series resistance.

High injection occurs in a forward biased p-n diode when the injected minority carrier density exceeds the doping density. High injection will therefore occur first in the lowest doped region of the diode since that region has the highest minority carrier density.

Using equations, one finds that high injection occurs in a p^+-n diode for the following applied voltage:

$$V_a = 2V_t In \frac{N_d}{n_i}$$

or at $V_a = 0.55$ V for the diode of Figure as can be verified on the figure as the voltage where the ideality factor changes from one to two. For higher forward bias voltages, the current no longer increases exponentially with voltage. Instead, it increases linearly due to the series resistance of the diode. This series resistance can be due to the contact resistance between the metal and the semiconductor, due to the resistivity of the semiconductor or due to the series resistance of the connecting wires. This series resistance increases the external voltage, V_a^*, relative to the internal voltage, V_a, considered so far.

$$V_a^* = V_a + IR_s$$

Where I is the diode current and R_s is the series resistance.

These four regions can be observed in most p-n diodes although the high-injection region rarely occurs, as the series resistance tends to limit the current first.

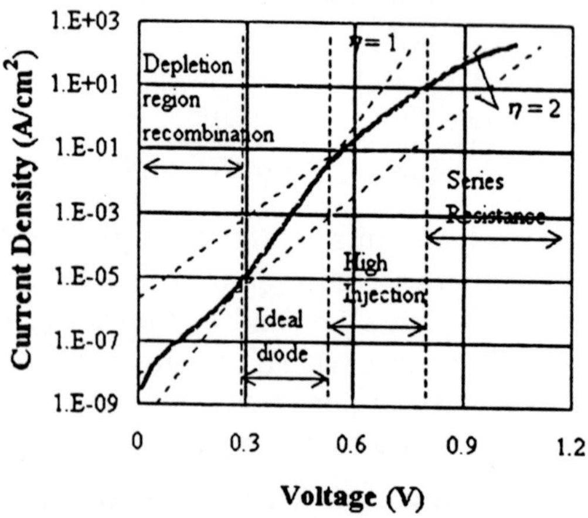

Fig. Current-Voltage characteristics of a silicon diode under forward bias.

THE DIFFUSION CAPACITANCE

As a p-n diode is forward biased, the minority carrier distribution in the quasi-neutral region increases dramatically.

In addition, to preserve quasi-neutrality, the majority carrier density increases by the same amount. This effect leads to an additional capacitance called the diffusion capacitance.

The diffusion capacitance is calculated from the change in charge with voltage:

$$C = \frac{d\Delta Q}{dV_a}$$

Where the charge, ΔQ, is due to the excess carriers. Unlike a parallel plate capacitor, the positive and negative charge is not spatially separated. Instead, the electrons and holes are separated by the energy bandgap. Nevertheless, these voltage dependent charges yield a capacitance just as the one associated with a parallel plate capacitor. The excess minority-carrier charge is obtained by integrating the charge density over the quasi-neutral region:

$$\Delta Q_p = \int_{x_n}^{w_n} qA(p_n - p_{n0})\, dx$$

We now distinguish between the two limiting cases as discussed when calculating the ideal diode current, namely the "long" diode and a "short" diode. The carrier distribution, $p_n(x)$, in a "long" diode is illustrated by Figure.

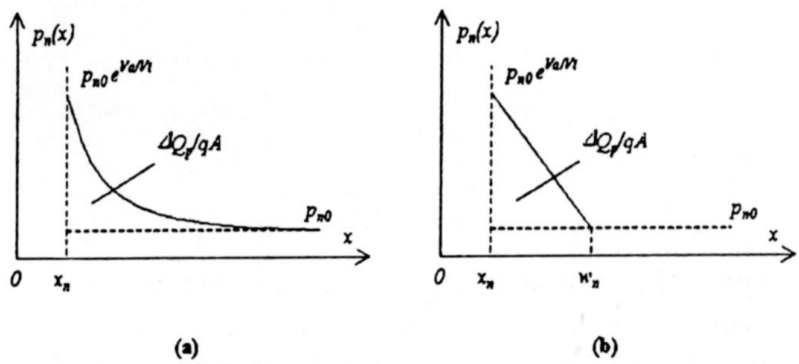

Fig. Minority carrier distribution in (a) a "long" diode, and (b) a "short" diode. The excess minority-carrier charge, ΔQ_p, in the quasi-neutral region, is proportional to the area defined by the solid and dotted lines.

Using equation, the excess charge, ΔQ_p, becomes:

$$\Delta Q_p = q A p_{n0}(e^{V_a/V_t} - 1)L_p = I_{s,p}(e^{V_a/V_t} - 1)\tau_p$$

where $I_{s,p}$ is the saturation current for holes, given by:

$$I_{s,p} = q\frac{A p_{n0} D_p}{L_p}$$

Equation directly links the excess charge to the diffusion current. Since all injected minority carriers recombine in the quasi-neutral region, the current equals the excess charge divided by the average time needed to recombine with the majority carriers, i.e. the carrier lifetime, τ_p. This relation will be the corner stone of the charge control model of bipolar junction transistors.

The diffusion capacitance then equals:

$$C_{d,p} = \frac{d(I_{s,p}(e^{V_a/V_t} - 1)\tau_p)}{dV_a} = \frac{I_{s,p}e^{V_a/V_t}\tau_p}{V_t}$$

Similarly, for a "short" diode, as illustrated by Figure, one obtains:

$$C_{d,p} = \frac{I_{s,p}e^{V_a/V_t}t_{r,p}}{V_t}$$

Where $t_{r,p}$ is the hole transit time given by:

$$t_{r,p} = \frac{w_n'^2}{2D_p}$$

Again, the excess charge can be related to the current. However, in the case of a "short" diode all minority carriers flow through the quasi-neutral region and do not recombine with the majority carriers. The current therefore equals the excess charge divided by the average time needed to traverse the quasi-neutral region, i.e. the transit time, $t_{r,p}$.

The total diffusion capacitance is obtained by adding the diffusion capacitance of the *n*-type quasi-neutral region to that of the *p*-type quasi-neutral region.

The total capacitance of the junction equals the sum of the junction capacitance, discussed in section 4.3.4, and the diffusion capacitance. For reverse biased voltages and small forward bias voltages, one finds that the junction capacitance is dominant. As the forward bias voltage is further increased the diffusion capacitance increases exponentially and eventually becomes larger than the junction capacitance.

Example

a. Calculate the diffusion capacitance of the diode described in Example at zero bias. Use m_n= 1000 cm^2/V-s, mp = 300 cm^2/V-s, w_p' = 1 mm and w_n' = 1 mm. The minority carrier lifetime equals 0.1 ms.

b. For the same diode, find the voltage for which the junction capacitance equals the diffusion capacitance.

Solution

a. The diffusion capacitance at zero volts equals

$$C_{d,0} = \frac{I_{s,p}\tau_p}{V_t} + \frac{I_{s,n}t_{r,n}}{V_t} = 1.73\times10^{-19}\ \text{F}$$

using

$$I_{s,p} = q\frac{A p_{n0} D_p}{L_p}$$

and

$$I_{s,n} = q\frac{A n_{p0} D_n}{w_p'}$$

where the "short" diode expression was used for the capacitance associated with the excess charge due to electrons in the p-type region. The "long" diode expression was used for the capacitance associated with the excess charge due to holes in the n-type region.The diffusion constants and diffusion lengths equal

$D_n = \mu_n \times V_t = 25.8\ \text{cm}^2/\text{s}$

$D_p = \mu_p \times V_t = 7.75\ \text{cm}^2/\text{s}$

$$L_p = \sqrt{D_p \tau_p}$$

And the electron transit time in the p-type region equals

$$t_{r,n} = \frac{w_p'^{2}}{2D_n} = 193\text{ps}$$

b. The voltage at which the junction capacitance equals the diffusion capacitance is obtained by solving

$$\frac{C_{j0}}{\sqrt{1 - \frac{V_a}{\phi_i}}} = C_{d,0} e^{V_a / V_t}$$

yielding $V_a = 0.442$ V

HIGH INJECTION EFFECTS

High injection of carriers causes to invalidate one of the approximations made in the derivation of the ideal diode characteristics, namely that the majority carrier density equals the thermal equilibrium value. Excess carriers will dominate the electron and hole concentration and can be expressed in the following way:

$$n_p p_p = n_i^2 \exp(\frac{V_a}{V_t}) = n_p (p_{p0} + n_p)$$

$$n_n p_n = n_i^2 \exp(\frac{V_a}{V_t}) = p_n (n_{n0} + p_n)$$

where all carrier densities with subscript n are taken at $x = x_n$ and those with subscript p at $x = -x_p$. Solving the resulting quadratic equation yields:

$$n_p = \frac{N_a}{2}\left(\sqrt{1+\frac{4n_i^2\exp(\frac{V_a}{V_t})}{N_a^2}}-1\right) \cong n_i^2\exp(\frac{V_a}{2V_t})$$

$$p_n = \frac{N_d}{2}\left(\sqrt{1+\frac{4n_i^2\exp(\frac{V_a}{V_t})}{N_d^2}}-1\right) \cong n_i^2\exp(\frac{V_a}{2V_t})$$

where the second terms are approximations for large V_a. From these expressions one can calculate the minority carrier diffusion current assuming a "long" diode in both quasi-neutral regions. We also ignore carrier recombination in the depletion region.

$$J_n + J_p = q\left(\frac{D_n}{L_n}+\frac{D_p}{L_p}\right)n_i\exp(\frac{V_a}{2V_t})$$

This means that high injection in a p-n diode will reduce the slope on the current-voltage characteristic on a semi-logarithmic scale to 119mV/decade.

High injection also causes a voltage drop across the quasi-neutral region. This voltage can be calculated from the carrier densities. Let's assume that high injection only occurs in the (lower doped) *p*-type region. The hole density at the edge of the depletion region ($x = x_p$) equals:

$$p_p(x = x_p) = N_a\exp(\frac{V_1}{V_t}) = N_a + n_p(x = x_p) = N_a + n_{p0}\exp(\frac{V_a - V_1}{V_t})$$

where V_1 is the voltage drop across the *p*-type quasi-neutral region. This equation can then be solved for V_1 yielding

$$V_1 = V_t\ln\left[\frac{\sqrt{1+\frac{4n_i^2\exp(\frac{V_a}{V_t})}{N_a^2}}-1}{2}\right]$$

High injection occurs (by definition) when the excess minority carrier density exceeds the doping density in the material.

It is under such conditions that also the majority carrier density increases since, for charge neutrality to exist, the excess electron density has to equal the excess hole density: If there exists a net charge, the resulting electric field causes the carriers to move so that charge neutrality is restored.

We repeat the high injection analysis without assuming a "large" applied voltage V_a, while providing more detail and deriving equations.

The analysis starts by assuming a certain excess carrier density so that the total density can be written as the sum of the thermal equilibrium density plus the excess carrier density. For an n-type region this yields the following equations:

$$n_n = n_n 0 + \delta n_n$$

$$p_n = p_{p0} + \delta p_n$$

The product of the carrier densities can be expressed as a function of the intrinsic density in the following way:

$$n_n p_n = n_i^2 \exp\frac{V_a}{V_t} = (n_{n0} + \delta n_n)(p_{p0} + \delta p_n)$$

where it was assumed that the semiconductor is non-degenerate and that the difference between the electron and hole quasi Fermi energies in electron volt equals the applied voltage in volt.. Quasi-neutrality implies that the excess densities are the same, which yields a quadratic equation for the minority carrier density p_n:

$$(\delta p_n)^2 + N_d \delta p_n - n_i^2\left(\exp\frac{V_a}{V_t} - 1\right) = 0$$

which yields a value for the minority carrier density at the edge of the depletion region.

$$\delta p_n = \frac{N_d}{2}\left(\sqrt{1+\frac{4n_i^2[\exp(\frac{V_a}{V_t})-1]}{N_d^2}}-1\right)$$

The associated current is a diffusion current and using a procedure similar to that for calculating the ideal diode current in a "long" diode one obtains the following hole current.

$$I_p = qA\frac{D_p\,\delta p_n}{L_p} = qA\frac{D_p N_d}{2L_p}\left(\sqrt{1+\frac{4n_i^2[\exp(\frac{V_a}{V_t})-1]}{N_d^2}}-1\right)$$

The electron current due to diffusion of electrons in the *p*-type region is given by a similar expression:

$$I_n = qA\frac{D_n\,\delta n_p}{L_n} = qA\frac{D_n N_a}{2L_n}\left(\sqrt{1+\frac{4n_i^2[\exp(\frac{V_a}{V_t})-1]}{N_a^2}}-1\right)$$

These expressions can be reduced to the ideal diode expressions provided that the excess minority density is much smaller than one quarter of the doping density, or:

$$I_p = qA\frac{D_p n_{p0}}{L_p}[\exp(\frac{V_a}{V_t})-1],\ \text{for}\ n_{p0}[\exp(\frac{V_a}{V_t})-1] << \frac{N_d}{4}$$

while if the excess minority carrier density is much larger than one quarter of the doping density and an expression is obtained which is only valid under high injection conditions:

$$I_p = qA\frac{D_p n_i}{L_p}\exp(\frac{V_a}{2V_t})\ ,\ \text{for}\ n_{p0}[\exp(\frac{V_a}{V_t})-1] >> \frac{N_d}{4}$$

A closer examination of the problem prompts the question whether the full depletion approximation is still valid since the sign of potential across the semiconductor reverses. However the increase of the majority carrier density beyond the doping density causes a potential variation across the "quasi-neutral" region. This voltage in the n-type region is given by:

$$V_n = V_t = \ln \frac{n_{n0} + \delta n_n}{n_{n0}} = V_t \ln\left[\frac{\sqrt{1 + \frac{4n_i^2 \exp(\frac{V_a}{V_t})}{N_d^2}} - 1}{2} \right]$$

and similarly for the p-type region:

$$V_p = V_t = \ln \frac{p_{p0} + \delta p_p}{p_{p0}} = V_t \ln\left[\frac{\sqrt{1 + \frac{4n_i^2 \exp(\frac{V_a}{V_t})}{N_a^2}} - 1}{2} \right]$$

The potentials across the "quasi-neutral" region causes a larger potential across the depletion layer, since they have opposite sign, so that the depletion layer width as calculated using the modified potential $\phi = \phi_i - V_a + V_n + V_p$ does not become zero.

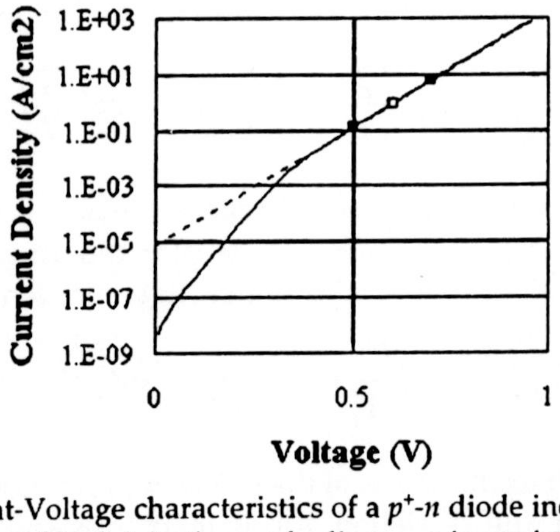

Fig. Current-Voltage characteristics of a p^+-n diode including the effects of high injection and a linear series resistance.

As an example we now consider an abrupt one-sided p-n diode. The current is shown as function of the voltage in Figure. It is calculated for a rather low *n*-type doping of 10^{13} cm^{-3} to further highlight the high injection effects.

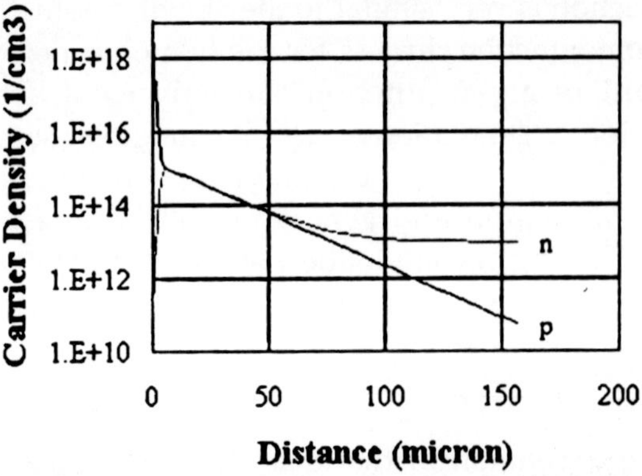

Fig. Electron and hole density under high injection conditions.

The dotted line on Figure fits the current at high forward bias. The slope is 1 decade/120 mV, which corresponds to an ideality factor of 2. Figure provides a more detailed look at the high injection condition. Figure presents the carrier densities in the *n*-type region. Once can observed that the majority carrier (electron) density increases beyond the doping density and tracks the minority carrier (hole) density in the region up to 50 mm away from the junction.

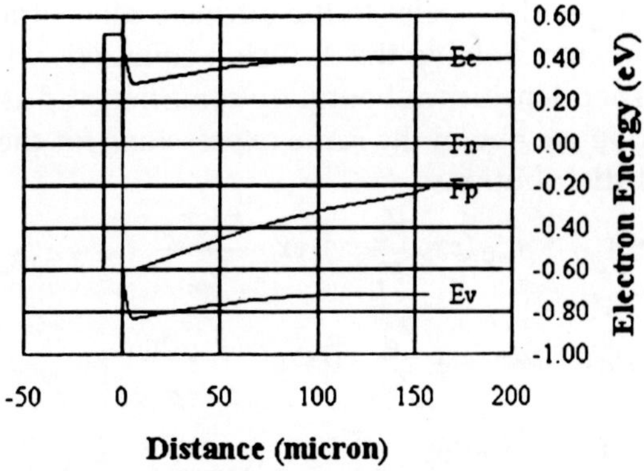

Fig. Energy band diagram under high injection conditions.

Figure provides the energy band diagram at a 0.6 V bias. Band bending can be observed in the quasi-neutral region.

HETEROJUNCTION DIODE CURRENT

This section is very similar to the one discussing currents across a homojunction. Just as for the homojunction we find that current in a p-n junction can only exist if there is recombination or generation of electron and holes somewhere throughout the structure. The ideal diode equation is a result of the recombination and generation in the quasi-neutral regions (including recombination at the contacts) whereas recombination and generation in the depletion region yield enhanced leakage or photocurrents.

Ideal Diode Equation

For the derivation of the ideal diode equation we will again assume that the quasi-Fermi levels are constant throughout the depletion region so that the minority carrier densities at the edges of the depletion region and assuming "low" injection are still given by:

$$n_p(x = x_p) = n_n \exp\left(-\frac{\phi_i - V_a}{V_t}\right) = \frac{n_{i,p}^2}{N_a} \exp\frac{V_a}{V_t}$$

$$p_n(x = -x_n) = p_p \exp\left(-\frac{\phi_i - V_a}{V_t}\right) = \frac{n_{i,n}^2}{N_d} \exp\frac{V_a}{V_t}$$

Where $n_{i,n}$ and $n_{i,p}$ refer to the intrinsic concentrations of the n and p region. Solving the diffusion equations with these minority carrier densities as boundary condition and assuming a "long" diode we obtain the same expressions for the carrier and current distributions:

$$n_p(x) = n_{p0} + n_{p0}\left(\exp\frac{V_a}{V_t} - 1\right)\exp\frac{x + x_p}{L_n} \quad \text{for } x < -x_p$$

$$p_n(x) = p_{n0} + p_{n0}\left(\exp\frac{V_a}{V_t} - 1\right)\exp\left(-\frac{x - x_n}{L_p}\right) \quad \text{for } x_n < x$$

$$J_n(x) = qD_n\frac{dn}{dx} = q\frac{D_n n_{p0}}{L_n}\left(\exp\frac{V_a}{V_t} - 1\right)\exp\frac{x + x_p}{L_n} \quad \text{for } x < -x_p$$

$$J_p(x) = qD_p \frac{dp}{dx} = q \frac{D_p p_{n0}}{L_p} (\exp \frac{V_a}{V_t} - 1) \exp(-\frac{x - x_n}{L_n}) \text{ for } x_n < x$$

Where $L_p L_n$ are the hole respectively the electron diffusion lengths in the n-type and p-type material, respectively. The difference compared to the homojunction case is contained in the difference of the material parameters, the thermal equilibrium carrier densities and the width of the depletion layers. Ignoring recombination of carriers in the base yields the total ideal diode current density J_{ideal}:

$$J_{ideal} = J_n(x = x_p) + J_p(x = -x_n) = q(\frac{D_n n_{p0}}{L_n} + \frac{D_p p_{n0}}{L_p})(\exp \frac{V_a}{V_t} - 1)$$

$$J_{ideal} = q(\frac{D_n n_{i,p}^2}{L_n N_a} + \frac{D_p n_{i,n}^2}{L_p N_d})(\exp \frac{V_a}{V_t} - 1)$$

This expression is valid only for a p-n diode with infinitely long quasi-neutral regions. For diodes with a quasi-neutral region shorter than the diffusion length, and assuming an infinite recombination velocity at the contacts, the diffusion length can simply be replaced by the width of the quasi-neutral region. For more general boundary conditions.

Since the intrinsic concentrations depend exponentially on the energy bandgap, a small difference in bandgapbetween the *n*-type and *p*-type material can cause a significant difference between the electron and hole current and that independent of the doping concentrations.

Recombination/generation in the Depletion Region

Recombination/generation currents in a heterojunction can be much more important than in a homojunction because most recombination/generation mechanisms depend on the intrinsic carrier concentration which depends strongly on the energy bandgap. We will consider only two major mechanisms: band-to-band recombination and Shockley-Hall-Read recombination.

Band-to-band Recombination

The recombination/generation rate is due to band-to-band transitions is given by:

$$U_{b-b} = b\left(np - n_i^2\right)$$

where b is the bimolecular recombination rate. For bulk GaAs this value is 1.1 x 10^{-10} cm^3s^{-1}. For $np > n_i^2$ (or under forward bias conditions) recombination dominates, whereas for $np < n_i^2$ (under reverse bias conditions) thermal generation of electron-hole pairs occurs. Assuming constant quasi-Fermi levels in the depletion region, this rate can be expressed as a function of the applied voltage by using the "modified" mass-action law $np = n_i^2 e^{V_a/V_t}$,yielding:

$$U_{b-b} = bn_i^2 (\exp \frac{V_a}{V_t} - 1)$$

The current is then obtained by integrating the recombination rate throughout the depletion region:

$$J_{b-b} = q \int_{x_n}^{x_p} U_{b-b} dx$$

For uniform material (homojunction) this integration yielded:

$$J_{b-b} = qbn_i^2 (\exp \frac{V_a}{V_t} - 1)x_d$$

Whereas for a p-n heterojunction consisting of two uniformly doped regions with different bandgap, the integral becomes:

$$J_{b-b} = qb(n_{i,n}^2 x_n + n_{i,p}^2 x_p)(\exp \frac{V_a}{V_t} - 1)$$

Schockley-Hall-Read Recombination

Provided bias conditions are "close" to thermal equilibrium the recombination rate due to a density N_t of traps with energy E_t and a recombination/generation cross-section σ is given by

$$U_{SHR} = \frac{np - n_i^2}{n + p + 2n_i \cosh \frac{E_i - E_t}{kT}} N_t \sigma v_{th}$$

where n_I is the intrinsic carrier concentration, v_{th} is the thermal velocity of the carriers and E_I is the intrinsic energy level. For $E_I = E_t$ and $\tau_0 = 1/N_t\sigma v_{th}$ this expression simplifies to:

$$U_{SHR} = \frac{np - n_i^2}{n + p + 2n_i \cosh \frac{E_i - E_t}{kT}} \frac{1}{\tau_0}$$

Throughout the depletion region, the product of electron and hole density is given by the "modified" mass action law:

$$np = n_i^2 (\exp \frac{V_a}{2V_t} - 1)$$

This enables to find the maximum recombination rate which occurs for $n = p = ni^{e^{V_a/2V_t}}$

$$U_{SHR,\max} = \frac{n_i}{2\tau_0} (\exp \frac{V_a}{2V_t} - 1)$$

The total recombination current is obtained by integrating the recombination rate over the depletion layer width:

$$\Delta J_n = -\Delta J_p = q \int_{-x_p}^{x_n} U_{SHR} dx$$

which can be written as a function of the maximum recombination rate and an "effective" width x':

$$\Delta J_n = qU_{SHR,\max} x' = q\frac{x' n_i}{2\tau_0}(\exp\frac{V_a}{2V_t} - 1)$$

where

$$x' = \frac{\int_{-x_p}^{x_n} U_{SHR} dx}{U_{SHR,\max}}$$

Since $U_{SHR,max}$ is larger than or equal to U_{SHR} anywhere within the depletion layer one finds that x' has to be smaller than $x_d = x_n + x_p$. (Note that for a p-i-N or p-qw-N structure the width of the intrinsic/qw layer has to be included).

The calculation of x' requires a numerical integration. The carrier concentrations n and p in the depletion region are given by:

$$n = N_c \exp\frac{E_{F,n} - E_c}{kT}$$

$$p = N_v \exp\frac{E_v - E_{F,p}}{kT}$$

Substituting these equations into then yields x'.

Recombination/generation in a Quantum Well

Band-to-band Recombination

Recognizing that band-to-band recombination between different states in the quantum well has a different coefficient, the total recombination including all possible transitions can be written as:

$$U_{b-b,qw} = B_1(N_1 P_1 - N_{i,1}^2) + B_2(N_2 P_2 - N_{i,2}^2) + \ldots$$

with

$$N_{i,n}^2 = N_{c,qw} N_{v,qw} \exp(-\frac{E_{g,qw,n}}{kT})$$

and

$$E_{g,qw,n} = E_g + E_{n,e} + E_{n,h}$$

where $E_{n,e}$ and $E_{n,h}$ are calculated in the absence of an electric field. To keep this derivation simple, we will only consider radiative transitions between the n = 1 states for which:

$$N_1 = N_{c,qw} \ln(1 + \exp \frac{E_{F,n} - E_c - E_{1,e}}{kT})$$

$$P_1 = N_{v,qw} \ln(1 + \exp \frac{E_v - E_{F,p} - E_{1,h}}{kT})$$

both expressions can be combined yielding

$$V_a = \frac{E_{F,n} - E_{F,p}}{q} = V_t \ln[(e^{N_1 / N_{c,qw}} - 1)(e^{P_1 / N_{v,qw}} - 1)] + \frac{E_{g,qw,1}}{q}$$

SHR Recombination

A straightforward extension of the expression for bulk material to two dimensions yields

$$U_{SHR,qw} = \frac{NP - N_i^2}{N + P + 2N_i} \frac{1}{\tau_0}$$

and the recombination current equals:

$$\Delta J_n = -\Delta J_p = qU_{SHR,qw} = q \frac{NP - N_i^2}{N + P + 2N_i} \frac{1}{\tau_0}$$

This expression implies that carriers from any quantum state are equally likely to recombine with a midgap trap.

The Graded p-n diode

General Discussion of a Graded Region

Graded regions can often be found in heterojunction devices. Typically they are used to avoid abrupt heterostructures, which limit the current flow. In addition they are used in laser diodes where they provide a graded index region, which guides the lasing mode. An accurate solution for a graded region requires the solution of a set of non-linear differential equations.

Numeric simulation programmes provide such solutions and can be used to gain the understanding needed to obtain approximate analytical solutions. A common misconception regarding such structures is that the flatband diagram is close to the actual energy band diagram under forward bias. Both are shown in the figure below for a single-quantum-well graded-index separate-confinement heterostructure (GRINSCH) as used in edge-emitting laser diodes.

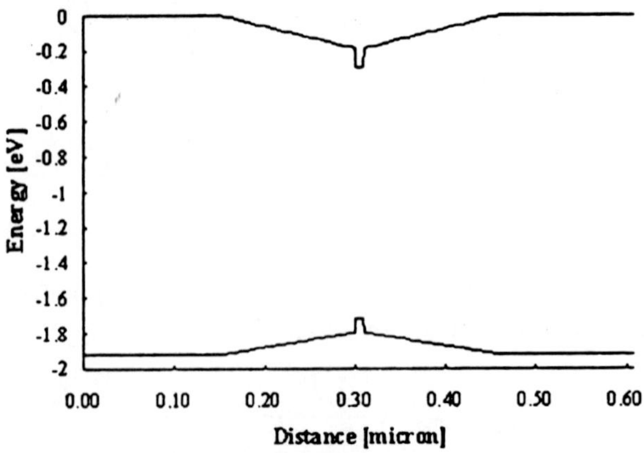

Fig. Flat band diagram of a graded AlGaAs p-n diode with x = 40 per cent in the cladding regions, x varying linearly from 40 per cent to 20 per cent in the graded regions and x = 0per cent in the quantum well.

The first difference between the two diagrams is that the conduction band edge in the n-type graded region as well as the valence band edge in the p-type graded region are almost constant. This assumption is correct if the majority carrier quasi-Fermi energy, the majority carrier density and the effective density of states for the majority carriers don't vary within the graded region. Since the carrier recombination primarily occurs within the quantum well (as it should be in a good laser diode), the quasi-Fermi energy does not change in the graded regions, while the effective density of states varies as the 3/2 power of the effective mass, which varies only slowly within the graded region.

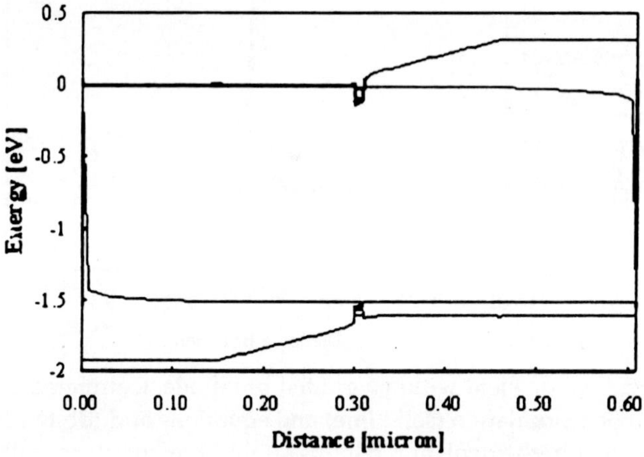

Fig. Energy band diagram of the graded p-n diode shown above under forward bias. V_a = 1.5 V, $N_a = 4 \times 10^{17}$ cm^{-3}, $N_d = 4 \times 10^{17}$ cm^{-3}. Shown are the conduction and valence band edges (solid lines) as well as the quasi-Fermi energies (dotted lines).

The constant band edge for the minority carriers implies that the minority carrier band edge reflects the bandgap variation within the graded region. It also implies a constant electric field throughout the grade region which compensates for the majority carrier bandgap variation or:

$$\mathcal{E}_{gr} = -\frac{1}{q}\frac{dE_{c0}(x)}{dx} \quad \text{in the } n\text{-type graded region}$$

$$\mathcal{E}_{gr} = -\frac{1}{q}\frac{dE_{v0}(x)}{dx} \quad \text{in the } p\text{-type graded region}$$

where $E_{c0}(x)$ and $E_{v0}(x)$ are the conduction and valence band edge as shown in the flatband diagram. The actual electric field is compared to these equations in Figure. The existence of an electric field requires a significant charge density at each end of the graded regions, which is obtained by a depletion of carriers. This also causes a small cusp in the band diagram.

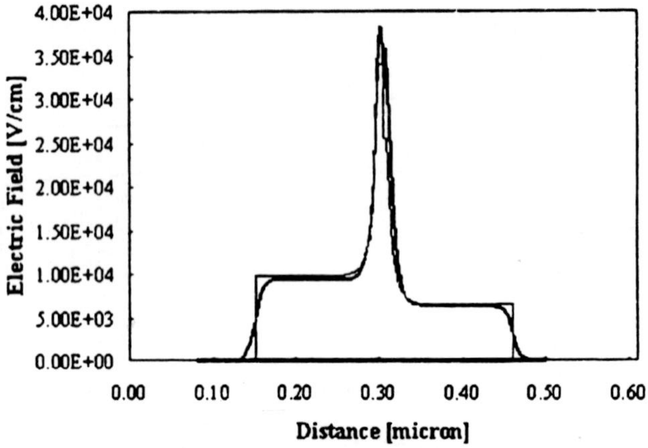

Fig. Electric Field within a graded p-n diode. Compared are a numeric simulation (solid line) and equations and (dotted line). The field in the depletion regions around the quantum well was calculated using the linearized Poisson equation.

Another important issue is that the traditional current equation with a drift and diffusion term must be modified. We now derive the modified expression by starting from the relation between the current density and the gradient of the quasi-Fermi level:

$$J_n = \mu_n n \frac{dF_n}{dx} = \mu_n n \frac{dE_c}{dx} + \mu_n n \frac{d(kT \ln \frac{n}{N_c})}{dx}$$

$$J_n = \mu_n n \frac{dF_n}{dx} = qD_n \frac{dn}{dx} - q\mu_n V_t \frac{n}{N_c} \frac{dN_c}{dx}$$

where it was assumed that the electron density is non-degenerate. At first sight it seems that only the last term is different from the usual expression. However the equation can be rewritten as a function of $E_{c0}(x)$, yielding:

$$J_n = q\mu_n n(\mathcal{E} + \frac{1}{q}\frac{dE_{c0}(x)}{dx} - \frac{V_t}{N_c}\frac{dN_c}{dx}) + qD_n \frac{dn}{dx}$$

This expression will be used in the next section to calculate the ideal diode current in a graded p-n diode. We will at that time ignore the gradient of the effective density of states. A similar expression can be derived for the hole current density, J_p.

We now calculate the ideal diode current in a graded heterojunction. Such calculation poses a special challenge since a gradient of the bandedge exists within the quasi-neutral region. The derivation below can be applied to a p-n diode with a graded doping density as well as one with a graded bandgap provided that the gradient is constant. For a diode with a graded doping concentration this implies an exponential doping profile as can be found in an ion-implanted base of a silicon bipolar junction transistor. For a diode with a graded bandgap the bandedge gradient is constant if the bandgap is linearly graded provided the majority carrier quasi-Fermi level is parallel to the majority carrier band edge.

Focusing on a diode with a graded bandgap we first assume that the gradient is indeed constant in the quasi-neutral region and that the doping density is constant. Using the full depletion approximation one can then solve for the depletion layer width. This requires solving a transcendental equation since the dielectric constant changes with material composition (and therefore also with bandgap energy). A first order approximation can be obtained by choosing an average dielectric constant within the depletion region and using previously derived expressions for the depletion layer width. Under forward bias conditions one finds that the potential across the depletion regions becomes comparable to the thermal voltage. One can then use the linearized Poisson equation or solve Poisson's equation exactly. The former approach was taken to obtain the electric field in Figure.

The next step requires solving the diffusion equation in the quasi-neutral region with the correct boundary condition and including the minority carrier bandedge gradient. For electrons in a *p*-type quasi-neutral region we have to solve the following modified diffusion equation.

$$0 = D_n \frac{d^2 n}{dx^2} + \mu_n \frac{1}{q} \frac{dE_c}{dx} \frac{dn}{dx} - \frac{n}{\tau_n}$$

which can be normalized yielding:

$$0 = n'' + 2\alpha n' - \frac{n}{L_n^2}$$

with $\alpha = \frac{1}{q2V_t} \frac{dE_c}{dx}$, $L_n^2 = D_n \tau_n$, $n' = \frac{dn}{dx}$ and $n'' = \frac{d^2 n}{dx^2}$

If the junction interface is at $x = 0$ and the p-type material is on the right hand side, extending up to infinity, the carrier concentrations equals

$$n(x) = n_{p0}(x_p) e^{V_a/V_t} \exp[-\alpha(1 + \sqrt{1 + \frac{1}{(L_n \alpha)^2}}(x - x_p)]$$

where we ignored the minority carrier density under thermal equilibrium, which limits this solution to forward bias voltages. Note that the minority carrier concentration $n_{p0}(x_p)$ at the edge of the depletion region (at $x = x_p$) is strongly voltage dependent since it is exponentially dependent on the actual bandgap at $x = x_p$.

The electron current at $x = x_p$ is calculated using the above carrier concentration but including the drift current since the bandedge gradient is not zero, yielding:

$$J_n = -qD_n n_{p0}(x_p) e^{V_a/V_t} \alpha(\sqrt{1 + \frac{1}{(L_n \alpha)^2}} - 1)$$

The minus sign occurs since the electrons move from left to right for a positive applied voltage. For $\alpha = 0$, the current equals the ideal diode current in a non-graded junction:

$$J_n = \frac{qD_n n_{p0}}{L_n} e^{V_a/V_t}$$

while for strongly graded diodes ($\alpha L_n >> 1$) the current becomes:

$$J_n = \frac{qD_n n_{p0}(x_n)}{2\alpha L_n^2} e^{V_a/V_t}$$

For a bandgap grading given by:

$$E_g = E_{g0} + \Delta E_g \frac{x}{d}$$

one finds

$$\alpha = \frac{\Delta E_g}{2kTd}$$

and the current density equals:

$$J_n = J_n(\alpha = 0) \frac{kT}{\Delta E_g} \frac{d}{L_n} \frac{n_{p0}(x_p)}{n_{p0}(0)}$$

where $J_n(\alpha = 0)$ is the current density in the absence of any bandgap grading.

REVERSE BIAS BREAKDOWN

GENERAL BREAKDOWN CHARACTERISTICS

The maximum reverse bias voltage that can be applied to a p-n diode is limited by breakdown. Breakdown is characterized by the rapid increase of the current under reverse bias. The corresponding applied voltage is referred to as the breakdown voltage.

The breakdown voltage is a key parameter of power devices. The breakdown of logic devices is equally important as one typically reduces the device dimensions without reducing the applied voltages, thereby increasing the internal electric field.

Two mechanisms can cause breakdown, namely avalanche multiplication and quantum mechanical tunneling of carriers through the bandgap. Neither of the two breakdown mechanisms is destructive. However heating caused by the large breakdown current and high breakdown voltage causes the diode to be destroyed unless sufficient heat sinking is provided.

Breakdown in silicon at room temperature can be predicted using the following empirical expression for the electric field at breakdown.

$$|\mathcal{E}_{br}| = \frac{4 \times 10^5}{1 - \frac{1}{3}\log(N/10^{16})} \text{ V/cm}$$

Assuming a one-sided abrupt p-n diode, the corresponding breakdown voltage can then be calculated, yielding:

$$|V_{br}| = -\phi_i + \frac{|\mathcal{E}_{br}|^2 \varepsilon_s}{2qN}$$

The resulting breakdown voltage is inversely proportional to the doping density if one ignores the weak doping dependence of the electric field at breakdown. The corresponding depletion layer width equals:

$$w_{br} = \frac{|\mathcal{E}_{br}| \varepsilon_s}{qN}$$

EDGE EFFECTS

Few p-n diodes are truly planar and typically have higher electric fields at the edges. Since the diodes will break down in the regions where the breakdown field is reached first, one has to take into account the radius of curvature of the metallurgical junction at the edges. Most doping processes including diffusion and ion implantation yield a radius of curvature on the order of the junction depth, x_j. The p-n diode interface can then be approximated as having a cylindrical shape along a straight edge and a spherical at a corner of a rectangular pattern. Both structures can be solved analytically as a function of the doping density, *N*, and the radius of curvature, x_j.

The resulting breakdown voltages and depletion layer widths are plotted below as a function of the doping density of an abrupt one-sided junction.

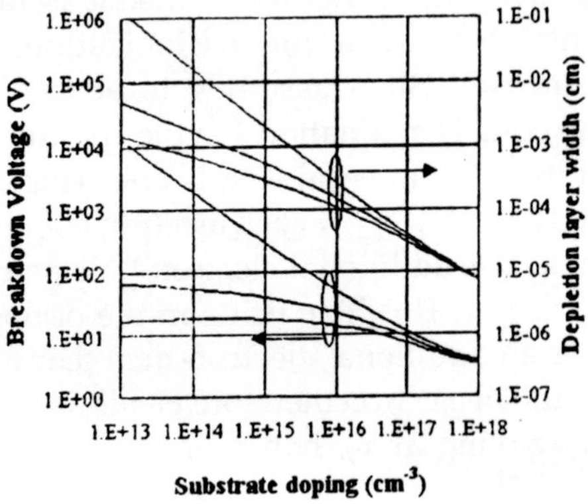

Fig. Breakdown voltage and depletion layer width at breakdown versus doping density of an abrupt one-sided p-n diode. Shown are the voltage and width for a planar (top curves), cylindrical (middle curves) and spherical (bottom curves) junction with 1 mm radius of curvature.

AVALANCHE BREAKDOWN

Avalanche breakdown is caused by impact ionization of electron-hole pairs. When applying a high electric field, carriers gain kinetic energy and generate additional electron-hole pairs through impact ionization. The ionization rate is quantified by the ionization constants of electrons and holes, α_n and α_p. These ionization constants are defined as the change of the carrier density with position divided by the carrier density or:

$$d_n = \alpha_n n dx$$

The ionization causes a generation of additional electrons and holes. Assuming that the ionization coefficients of electrons and holes are the same, the multiplication factor *M*, can be calculated from:

$$M = \frac{1}{1 - \int_{x_1}^{x_2} \alpha dx}$$

The integral is taken between x_1 and x_2, the region within the depletion layer where the electric field is assumed constant and large enough to cause impact ionization. Outside this range, the electric field is assumed to be too low to cause impact ionization. The equation for the multiplication factor reaches infinity if the integral equals one. This condition can be interpreted as follows: For each electron coming to the high field at point x_1 one additional electron-hole pair is generated arriving at point x_2. This hole drifts in the opposite direction and generates an additional electron-hole pair at the starting point x_1. One initial electron therefore yields an infinite number of electrons arriving at x_2, hence an infinite multiplication factor.

The multiplication factor is commonly expressed as a function of the applied voltage and the breakdown voltage using the following empirical relation:

$$M = \frac{1}{1 - \left|\frac{V_a}{V_{br}}\right|^n}, \text{ where } 2 < n < 6$$

ZENER BREAKDOWN

Quantum mechanical tunneling of carriers through the bandgap is the dominant breakdown mechanism for highly doped p-n junctions. The analysis is identical to that of tunneling in a metal-semiconductor junction where the barrier height is replaced by the energy bandgap of the material.

The tunneling probability equals:

$$\Theta = \exp\left(-\frac{4}{3}\frac{\sqrt{2m^*}}{q\hbar}\frac{E_g^{3/2}}{\mathcal{E}}\right)$$

where the electric field equals $\varepsilon = E_g/(qL)$.

The tunneling current is obtained from the product of the carrier charge, velocity and carrier density. The velocity equals the Richardson velocity, the velocity with which on average the carriers approach the barrier while the carrier density equals the density of available electrons multiplied with the tunneling probability, yielding:

$$J_n = q v_R n \Theta$$

The tunneling current therefore depends exponentially on the bandgap energy to the 3/2 power.

OPTOELECTRONIC DEVICES

P-n junctions are an integral part of several optoelectronic devices. These include photodiodes, solar cells light emitting diodes (LEDs) and semiconductor lasers. In this section, we discuss the principle of operation of these devices and derive equations for key parameters.

LIGHT ABSORPTION AND EMISSION

A large number of optoelectronic devices consist of a *p*-type and *n*-type region, just like a regular p-n diode. The key difference is that there is an additional interaction between the electrons and holes in the semiconductor and light. This interaction is not restricted to optoelectronic devices. Regular diodes are also known to be light sensitive and in some cases also emit light. The key difference is that optoelectronic devices such as photodiodes, solar cells, LEDs and laser diodes are specifically designed to optimize the light absorption and emission, resulting in a high conversion efficiency.

Light absorption and emission in a semiconductor is known to be heavily dependent on the detailed band structure of the semiconductor. Direct bandgap semiconductors, i.e semiconductors for which the minimum of the conduction band occurs at the same wavevector, *k*, as the maximum of the valence band, have a stronger absorption of light as characterized by a larger absorption coefficient. They are also the favored semiconductors when fabricating light emitting devices. Indirect bandgap semiconductors, i.e. semiconductors for which the minimum of the conduction band does not occur at the same wavevector as the maximum of the valence band, are known to have a smaller absorption coefficient and are rarely used in light emitting devices.

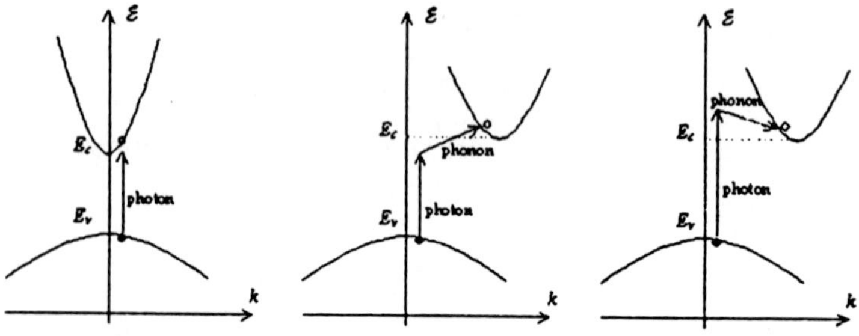

Fig. *E-k* diagram illustrating a) Photon absorption in a direct bandgap semiconductor b) Photon absorption in an indirect bandgap semiconductor assisted by phonon absorption and c) Photon absorption in an indirect bandgap semiconductor assisted by phonon emission.

This striking difference is further illustrated with Figure and can be explained based on the energy and momentum conservation required in the electron-photon interaction. The direct bandgap semiconductor, which has a vertically aligned conduction and valence band, is shown in Figure. Absorption of a photon is obtained if an empty state in the conduction band is available for which the energy and momentum equals that of an electron in the valence band plus that of the incident photon. Photons have little momentum relative of their energy since they travel at the speed of light. The electron therefore makes an almost vertical transition on the *E-k* diagram.

For an indirect bandgap semiconductor, the conduction band is not vertically aligned to the valence band as shown in Figure. Therefore a simply interaction of an incident photon with an electron in the valence band will not provide the correct energy and momentum corresponding to that of an empty state in the conduction band. As a result absorption of light requires the help of another particle, namely a photon. Since a phonon, i.e a particle associated with lattice vibrations, has a relatively low velocity close to the speed of sound in the material, it has a small energy and large momentum compared to that of a photon. Conservation of both energy and momentum can therefore be obtained in the absorption process if a phonon is created or an existing phonon participates. The phonon assisted absorption processes are illustrated with Figure. It illustrates the absorption of a photon aided by the simultaneous absorption of a phonon, while Figure depicts the absorption of a photon, which results in the emission of a phonon.

The minimum photon energy that can be absorbed is slightly below the bandgap energy in the case of phonon absorption and has to be slightly above the bandgap energy in the case of phonon emission. Since the absorption process in an indirect bandgap semiconductor involves a phonon in addition to the electron and photon, the probability of having an interaction take place involving all three particles will be lower than a simple electron-photon interaction in a direct bandgap semiconductor. As a result one finds that absorption is much stronger in a direct bandgap material.

Similarly, in the case of light emission, a direct bandgap material is also more likely to emit a photon than an indirect bandgap material. While indirect bandgap materials are occasionally used for some LEDs, they result in a low conversion efficiency. Direct bandgap materials are used exclusively for semiconductor laser diodes.

PHOTODIODES

Photodiodes and crystalline solar cells are essentially the same as the p-n diodes, which have been described in this chapter. However, the diode is exposed to light, which yields a photocurrent in addition to the diode current so that the total diode current is given by:

$$I = I_s (e^{V_a / V_t} - 1) - I_{ph}$$

where the additional photocurrent, I_{ph}, is due to photogeneration of electrons and holes shown in Figure These electrons and holes are pulled into the region where they are majority carriers by the electric field in the depletion region.

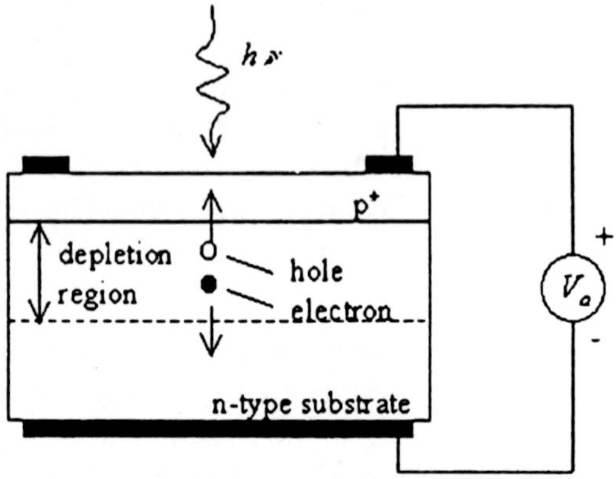

Fig. Motion of photo-generated carriers in a p-n photodiode.

The photo-generated carriers cause a photocurrent, which opposes the diode current under forward bias. Therefore, the diode can be used as a photodetector - using a reverse or even zero bias voltage - as the measured photocurrent is proportional to the incident light intensity. The diode can also be used as a solar cell - using a forward bias – to generate electrical power.

The primary characteristics of a photodiode are the responsivity, the dark current and the bandwidth. The responsivity is the photocurrent divided by the incident optical power. The maximum photocurrent in a photodiode equals

$$I_{ph,\max} = \frac{q}{h\zeta} P_{in}$$

Where P_{in} is the incident optical power. This maximum photocurrent occurs when each incoming photon creates one electron-hole pair, which contributes to the photocurrent. The maximum photocurrent in the presence of a reflection, R at the surface of the photodiode and an absorption over a thickness d, in a material with an absorption coefficient, α, is given by:

$$I_{ph} = (1-R)\left(1-e^{-ad}\right)\frac{qP_{in}}{h\zeta}$$

This photocurrent is obtained by integrating the generation rate over the absorption region with thickness, *d*. The photocurrent is further reduced if photo-generated electron-hole pairs recombine within the photodiode instead of being swept into the regions where they are majority carriers.

The dark current is the current through the diode in the absence of light. This current is due to the ideal diode current, the generation/recombination of carriers in the depletion region and any surface leakage, which occurs in the diode. The dark current obviously limits the minimum power detected by the photodiode, since a photocurrent much smaller than the dark current would be hard to measure.

However, the true limitation is the shot noise generated by the current through the diode. The shot noise as quantified by the average of the square of the noise current is given by:

$$\left\langle i^2 \right\rangle = 2qI\Delta f$$

Where I is the diode current and Δf is the bandwidth of the detector. The bandwidth of the diode is affected by the transit time of the photo-generated carriers through the diode and by the capacitance of the diode. The carrier transit time yields the intrinsic bandwidth of the diode while the capacitance together with the impedance of the amplifier or the transmission line connected to the diode yields a the parasitic RC delay.

SOLAR CELLS

Solar cells are typically illuminated with sunlight and are intended to convert the solar energy into electrical energy. The solar energy is in the form of electromagnetic radiation, more specifically "black-body" radiation.The sun's spectrum is consistent with that of a black body at a temperature of 5800 K. The radiation spectrum has a peak at 0.8 eV. A significant part of the spectrum is in the visible range of the spectrum (400 - 700 nm). The power density is approximately 100 mW/ cm^2.

Only part of the solar spectrum actually makes it to the earth's surface. Scattering and absorption in the earth's atmosphere, and the incident angle affect the incident power density. Therefore, the available power density depends on the time of the day, the season and the latitude of a specific location.

Of the solar light, which does reach a solar cell, only photons with energy larger than the energy bandgap of the semiconductor generate electron-hole pairs. In addition, one finds that the voltage across the solar cell at the point where it delivers its maximum power is less than the bandgap energy in electron volt. The overall power-conversion efficiency of single-crystalline solar cells ranges from 10 to 30 per cent yielding 10 to 30 mW/cm^2.

The calculation of the maximum power of a solar cell is illustrated by Figures. The sign convention of the current and voltage is shown as well. It considers a current coming out of the cell to be positive as it leads to electrical power generation. The power generated depends on the solar cell itself and the load connected to it. As an example, a resistive load is shown in the diagram below.

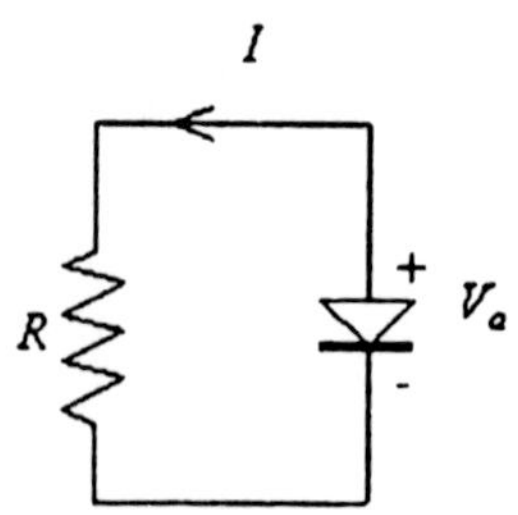

Fig. Circuit diagram and sign convention of a p-n diode solar cell connected to a resistive load.

The current and the power as function of the forward bias voltage across the diode are shown in Figure for a photocurrent of 1 mA:

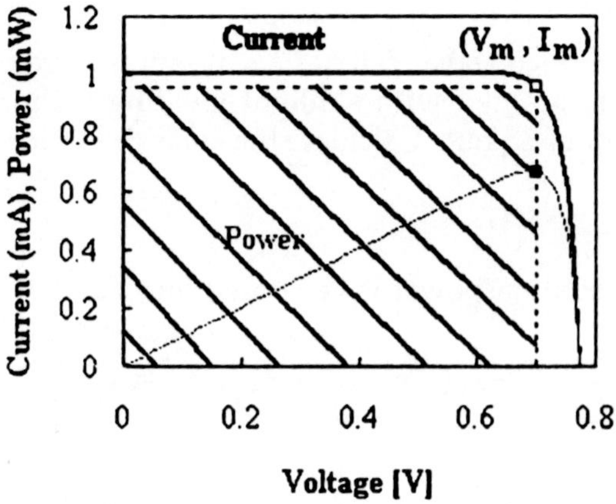

Fig. Current-Voltage (*I-V*) and Power-Voltage (*P-V*) characteristics of a p-n diode solar cell with I_{ph} = 1 mA and $I_s = 10^{-10}$ A. The crosshatched area indicates the power generated by the solar cell. The markers indicate the voltage and current, V_m and I_m, for which the maximum power, P_m is generated.

We identify the open-circuit voltage, V_{oc}, as the voltage across the illuminated cell at zero current. The *short-circuit* current, I_{sc}, is the current through the illuminated cell if the voltage across the cell is zero. The *short-circuit* current is close to the photocurrent while the *open-circuit* voltage is close to the turn-on voltage of the diode as measured on a current scale similar to that of the photocurrent.

The power equals the product of the diode voltage and current and at first increases linearly with the diode voltage but then rapidly goes to zero as the voltage approaches the turn-on voltage of the diode. The maximum power is obtained at a voltage labeled as V_m with I_m being the current at that voltage.

The fill factor of the solar cell is defined as the ratio of the maximum power of the cell to the product of the *open-circuit* voltage, V_{oc}, and the *short-circuit* current, I_{sc}, or:

$$\text{Fill Factor} = \frac{I_m V_m}{I_{sc} V_{oc}}$$

Example

A 1 cm^2 silicon solar cell has a saturation current of 10^{-12} A and is illuminated with sunlight yielding a short-circuit photocurrent of 25 mA. Calculate the solar cell efficiency and fill factor.

Solution

The maximum power is generated for:

$$\frac{dP}{dV_a} = 0 = I_s(e^{V_m/V_t} - 1) - I_{ph} + \frac{V_m}{V_t} I_s e^{V_m/V_t}$$

where the voltage, V_m, is the voltage corresponding to the maximum power point. This voltage is obtained by solving the following transcendental equation:

$$V_m = V_t \ln \frac{1 + I_{ph}/I_s}{1 + V_m/V_t}$$

Using iteration and a starting value of 0.5 V one obtains the following successive values for V_m:

V_m = 0.5, 0.542, 0.540 V

and the efficiency equals:

$$\eta = \left|\frac{V_m I_m}{P_{in}}\right| = \frac{0.54 \times 0.024}{0.1} = 13\%$$

The current, I_m, corresponding to the voltage, V_m, was calculated using equation and the power of the sun was assumed 100 mW/cm^2. The fill factor equals:

$$\text{fill factor} = \frac{V_m I_m}{V_{oc} I_{sc}} = \frac{0.54 \times 0.024}{0.62 \times 0.025} = 83\%$$

where the open circuit voltage is calculated using equation and I = 0. The short circuit current equals the photocurrent.

LEDS

Light emitting diodes are p-n diodes in which the recombination of electrons and holes yields a photon. This radiative recombination process occurs primarily in direct bandgap semiconductors where the lowest conduction band minimum and the highest valence band maximum occur at $k = 0$, where k is the wavenumber. Examples of direct bandgap semiconductors are GaAs, InP, and GaN while most group IV semiconductors including Si, Ge and SiC are indirect bandgap semiconductors.

The radiative recombination process is in competition with non-radiative recombination processes such as trap-assisted recombination. Radiative recombination dominates at high minority-carrier densities. Using a quantum well, a thin region with a lower bandgap, positioned at the metallurgical junction, one can obtain high carrier densities at low current densities. These quantum well LEDs have high internal quantum efficiency as almost every electron injected in the quantum well recombines with a hole and yields a photon.

The external quantum efficiency of planar LEDs is much lower than unity due to total internal reflection. As the photons are generated in the semiconductor, which has a high refractive index, only photons traveling normal to the semiconductor-air interface can exit the semiconductor. For GaAs with a refractive index of 3.5, the angle for total internal reflection equals 17° so that only a few percent of the generated photons can escape the semiconductor. This effect can be avoided by having a spherical semiconductor shape, which ensures that most photons travel normal to the interface. The external quantum efficiency can thereby be increased to values larger than 50per cent.

LASER DIODES

Laser diodes are very similar to LEDs since they also consist of a p-n diode with an active region where electrons and holes recombine resulting in light emission. However, a laser diode also contains an optical cavity where stimulated emission takes place. The laser cavity consists of a waveguide terminated on each end by a mirror. As an example, the structure of an edge-emitting laser diode is shown in Figure. Photons, which are emitted into the waveguide, can travel back and forth in this waveguide provided they are reflected at the mirrors.

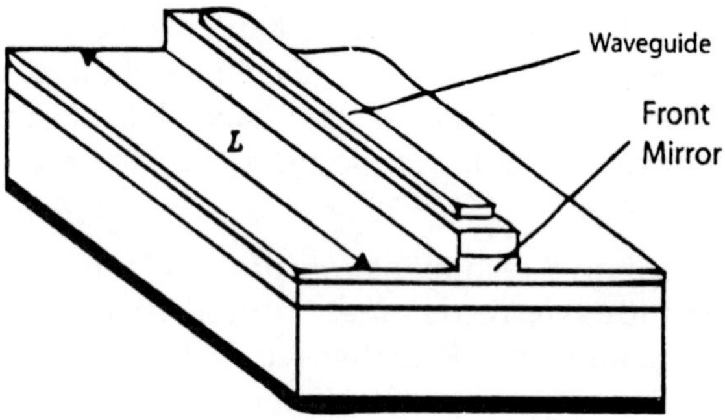

Fig. Structure of an edge-emitting laser diode.

The light in the waveguide is amplified by stimulated emission. Stimulated emission is a process where a photon triggers the radiative recombination of an electron and hole thereby creating an additional photon with the same energy and phase as the incident photon. This process is illustrated with Figure. This "cloning" of photons results in a coherent beam.

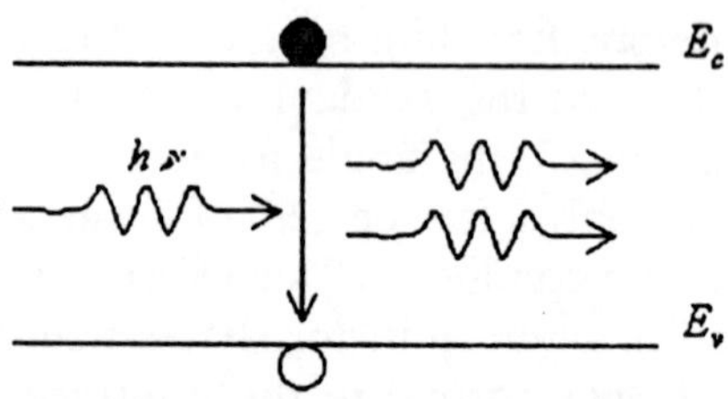

Fig. Stimulated emission of a photon.

The stimulated emission process yields an increase in photons as they travel along the waveguide. Combined with the waveguide losses, stimulated emission yields a net gain per unit length, *g*. The number of photons can therefore be maintained if the roundtrip amplification in a cavity of length, *L*, including the partial reflection at the mirrors with reflectivity R_1 and R_2 equals unity.

This yields the following lasing condition:

$$\text{Roundtrip amplification} = e^{2gL} R_1 R_2 = 1$$

If the roundtrip amplification is less than one, then the number of photons steadily decreases. If the roundtrip amplification is larger than one, the number of photons increases as the photons travel back and forth in the cavity. The gain required for lasing therefore equals:

$$g = \frac{1}{2L} \ln \frac{1}{R_1 R_2}$$

Initially, the gain is negative if no current is applied to the laser diode as absorption dominates in the waveguide. As the laser current is increased, the absorption first decreases and the gain increases. The current for which the gain satisfies the lasing condition is the threshold current of the laser, I_{th}. Below the threshold current very little light is emitted by the laser structure. For an applied current larger than the threshold current, the output power, P_{out}, increases linearly with the applied current, as each additional incoming electron-hole pair is converted into an additional photon. The output power therefore equals:

$$P_{out} = \eta \frac{h\nu}{q} (I - I_{th})$$

where $h\nu$ is the energy per photon. The factor, η, indicates that only a fraction of the generated photons contribute to the output power of the laser as photons are partially lost through the other mirror and throughout the waveguide.

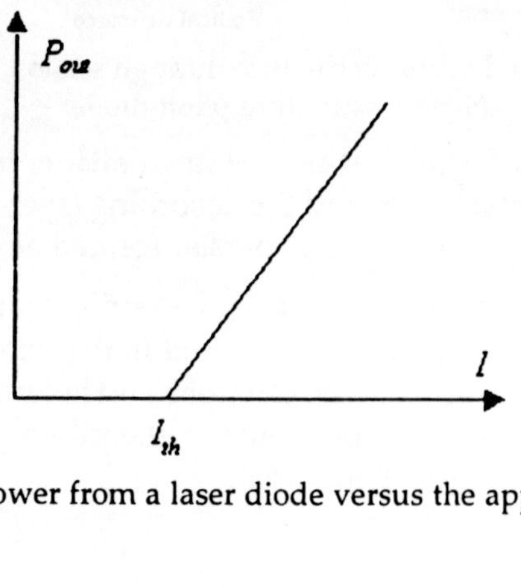

Fig. Output power from a laser diode versus the applied current.

PHOTODIODES

P-I-N PHOTODIODES

P-i-N photodiodes are commonly used in a variety of applications. A typical P-i-N photodiode is shown in Figure. It consists of a highly-doped transparent *p*-type contact layer on top of an undoped absorbing layer and an *n*-type highly doped contact layer on the bottom. Discrete photodiodes are fabricated on a conductive substrate as shown in the figure, which facilitates the formation of the *n*-type contact and reduces the number of process steps. The top contact is typically a metal ring contact, which has a low contact resistance and still allows the light to be absorbed in the semiconductor. An alternative approach uses a transparent conductor such as Indium Tin Oxide (ITO). The active device area is formed by mesa etching or by proton implantation of the adjacent area, which makes it isolating. A dielectric layer is added around the active area to reduce leakage currents and to ensure a low parasitic capacitance of the contact pad.

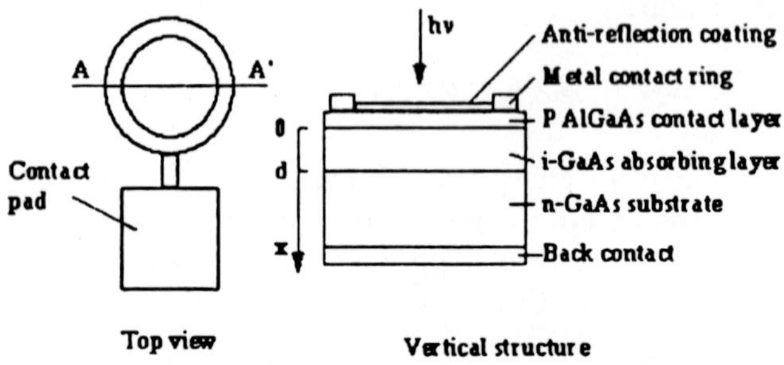

Fig. Top view and vertical structure through section A-A' of a P-i-N heterostructure photodiode.

Grading of the material composition between the transparent contact layer and the absorbing layer is commonly used to reduce the n-n^+ or p-p^+ barrier formed at the interface.

The above structure evolved mainly from one basic requirement: light should be absorbed in the depletion region of the diode to ensure that the electrons and holes are separated in the electric field and contribute to the photocurrent, while the transit time must be minimal.

This implies that a depletion region larger than the absorption length must exist in the detector. This is easily assured by making the absorbing layer undoped. Only a very small voltage is required to deplete the undoped region. If a minimum electric field is required throughout the absorbing layer, to ensure a short transit time, it is also the undoped structure, which satisfies this condition with a minimal voltage across the region, because the electric field is constant. An added advantage is that the recombination/generation time constant is longest for undoped material, which provides a minimal thermal generation current.

It also implies that the top contact layer should be transparent to the incoming light. In silicon photodiodes one uses a thin highly doped contact layer to minimize the absorption. By using a contact layer with a wider band gap (also called the window layer) absorption in the contact layer can be eliminated (except for a small fraction due to free carrier absorption) which improves the responsivity.

Electron-hole pairs, which are absorbed in the quasi-neutral regions, can still contribute to the photocurrent provided they are generated within one diffusion length of the depletion region.

However, the collection of carriers due to diffusion is relatively inefficient and leads to long tails in the transient response. It therefore should be avoided.

Because of the large difference in refractive index between air and most semiconductors, there is a substantial reflection at the surface. The reflection at normal incidence between two materials with refractive index n_1 and n_2 is given by:

$$R = \left(\frac{n_1 - n_2}{n_1 + n_2}\right)^2$$

For instance, the reflection between air and GaAs ($n = 3.5$) is 31 per cent.

By coating the semiconductor surface with a dielectric material (anti-reflection coating) of appropriate thickness this reflection can largely be eliminated.

The reflectivity for an arbitrary incident angle is:

$$R_{TE} = \left(\frac{n_1 \cos\theta_i - n_2 \cos\theta_t}{n_1 \cos\theta_i + n_2 \cos\theta_t}\right)^2$$

$$R_{TM} = \left(\frac{n_2 \cos\theta_i - n_1 \cos\theta_t}{n_2 \cos\theta_i + n_1 \cos\theta_t}\right)^2$$

with $n_2 \sin\theta_t = n_t \sin\theta_i$

where θ_i is the incident angle, and θ_t the transmitted angle. R_{TE} is the reflectivity if the electric field is parallel to the surface while R_{TM} is the reflectivity if the magnetic field is parallel to the surface. The reflectivity as a function of θ_i, for an air-GaAs interface is shown in Figure:

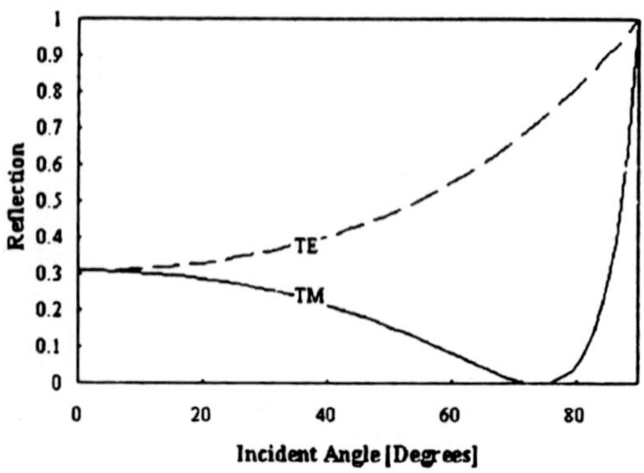

Fig. Reflectivity versus incident angle for a transverse electric, R_{TE}, and transverse magnetic, R_{TM}, incident field.

Responsivity of a P-i-N photodiode

Generation of electron hole pairs

The generation of electron-hole pairs in a semiconductor is directly related to the absorption of light since every absorbed photon generates one electron-hole pair. The optical generation rate g_{op} is given by:

$$g_{op} = -\frac{1}{A}\frac{dP_{opt}}{dx}\frac{1}{h\nu} = \frac{\alpha P_{opt}}{Ah\nu}$$

where A is the illuminated area of the photodiode, P_{opt} is the incident optical power, α is the absorption coefficient and $h\nu$ is the photon energy. Note that the optical power is position dependent and obtained by solving:

$$\frac{dP_{opt}}{dx} = -\alpha P_{opt}$$

The resulting generation rate must be added to the continuity equation and solved throughout the photodiode, which results in the photocurrent.

Photocurrent due to Absorption in the Depletion Region

Assuming that all the generated electron-hole pairs contribute to the photocurrent, the photocurrent is simply the integral of the generation rate over the depletion region:

$$I_{ph} = -qA \int_{-x_p}^{x_n+d} g_{op}\,dx$$

where d is the thickness of the undoped region. The minus sign is due to the sign convention indicated on Figure. For a P-i-N diode with heavily doped n-type and p-type regions and a transparent top contact layer, this integral reduces to:

$$I_{ph} = -\frac{q(1-R)P_{in}}{h\nu}(1-e^{-\alpha d})$$

where P_{in} is the incident optical power and R is the reflection at the surface.

Photocurrent Due to Absorption in the Quasi-neutral Region

To find the photocurrent due to absorption in the quasi-neutral region, we first have to solve the diffusion equation in the presence of light. For holes in the n-type contact layer this means solving the continuity equation:

$$\frac{\partial p_n}{dt} = -\frac{1}{q}\frac{\partial J_p}{dx} + \frac{p_n - p_{n0}}{\tau_p} + g_{op}(x)$$

Where the electron-hole pair generation g_{op} depends on position. For an the n-type contact layer with the same energy bandgap as the absorption layer, the optical generation rate equals:

$$g_{op}(x) = \frac{P_{in}(1-R)\,\alpha e^{-\alpha x}}{Ah\nu}$$

and the photocurrent due to holes originating in the n-type contact layer equals:

$$I_{ph} = -\frac{q(1-R)P_{in}e^{-\alpha d}}{h\nu}\frac{\alpha L_p}{1+\alpha L_p}(1-e^{-\alpha d}) - \frac{qD_p p_{n0}}{L_p}$$

The first term is due to light whereas the second term is the due to thermal generation of electron-hole pairs. This derivation assumes that the thickness of the n-type contact layer is much larger than the diffusion length.

Absorption in the p-contact Region

Even though the contact layer was designed so that no light absorbs in this layer, it will become absorbing at shorter wavelengths. Consider a worst-case scenario where all the electron-hole pairs, which are generated in the p-type contact layer, recombine without contributing to the photocurrent. The optical power incident on the undoped region is reduced by $\exp(-\alpha^* w_p')$ where w_p' is the width of the quasi-neutral p region and α^* is the absorption coefficient in that region.

Total Responsivity:

Combining all the above effects the total responsivity of the detector - ignoring the dark current - equals:

$$\mathcal{R} = \left|\frac{I_{ph}}{P_{in}}\right| = \frac{q(1-R)\exp(-\alpha^* w_p')}{h\nu}[1-\frac{e^{-\alpha d}}{1+\alpha L}]$$

Note that α^*, α and $h\nu$ are wavelength dependent. For a direct bandgap semiconductor these are calculated from:

$$\alpha = K\sqrt{E_{ph} - E_g} \text{ and } \alpha^* = K^*\sqrt{E_{ph} - E_g^*}$$

The quantum efficiency then equals:

$$n = \frac{Rh\nu}{q} = (1 - R)\exp\left(-a^* w' p\right)\left[1 - \frac{e^{-\alpha d}}{1 + \alpha L}\right]$$

Dark Current of the Photodiode:

The dark current of a p-n diode including the ideal diode current, as well as recombination/generation in the depletion region is given by:

$$I_{p-i-n} = qA\left(\frac{D_p p_{n0}}{L_p} + \frac{D_n n_{p0}}{w_p'} + bdn_{i,u}^2\right)\left(e^{V_a/V_t} - 1\right)$$

$$+ qA\frac{x' n_{i,u}}{2\zeta_{nr}}\left(e^{V_a/2V_t} - 1\right)$$

Under reverse bias conditions this expression reduces to:

$$I_{p-i-n} = -qA\left[\frac{D_p p_{n0}}{L_p} + bdn_{i,u}^2 + \frac{x' n_{i,u}}{2\zeta_{nr}}\right]$$

The ideal diode current due to recombination of electrons has been ignored since $n_{p0} = n_{i,p}^2/N_a$ is much smaller than p_{n0} because the *p*-layer has a larger band gap. In the undoped region, one expects the trap-assisted generation to be much larger than bimolecular generation. Which further reduces the current to:

$$I_{p-i-n} = -qA\left[\frac{D_p n_i^2, n}{L_p N_d} + \frac{x' n_i, u}{2\zeta_{nr}}\right]$$

The trap-assisted recombination tends to dominate for most practical diodes.

Noise in a Photodiode

Shot Noise Sensitivity

Noise in a p-i-n photodiode is primarily due to shot noise; the random nature of the generation of carriers in the photodiode yields also a random current fluctuation. The square of the current fluctuations equals:

$$\left\langle i^2 \right\rangle = 2q\left(\sum_j I_j\right)\Delta f$$

where I_j are the currents due to different recombination/ generation mechanisms and Δf is the frequency range. Including the ideal diode current, Shockley-Hall-Read and band-to-band recombination as well as generation due to light one obtains:

$$\langle i^2 \rangle = 2q\Delta f\{qA[\frac{D_n n_{p0}}{w_p'} + \frac{D_p p_{n0}}{L_p} + bdn_{i,u}^2 + \frac{x' n_{i,u}}{2\tau_{nr}}]$$

$$+ \frac{qAn_i x'}{2\tau_{nr}}(1 + e^{V_a/2V_t}) + P_{in}\mathcal{R}\}$$

The minimum detectable input power depends on the actual signal and the required signal to noise ration. As a first approximation, we now calculate the minimum detectable power as the power, which generates a current equal to the RMS noise current. A more detailed model for sinusoidal modulated signals.

$$P_{\min} = \frac{\sqrt{\left\langle i^2 \right\rangle}}{R}$$

The minimal noise current is obtained at $V_a = 0$ for which the noise current and minimal power equal:

$$\langle i^2 \rangle = 2q\ \Delta f[2\frac{n_{i,n}^2 D_p}{L_p N_d} + 2\frac{n_{i,u} x'}{2\tau_{nr}} + P_{in}\mathcal{R}]$$

$$P_{min} = \sqrt{\frac{2q\,\Delta f}{\mathcal{R}^2}[2\frac{n_{i,n}^2 D_p}{L_p N_d} + 2\frac{n_{i,u} x'}{2\tau_{nr}} + P_{in}\mathcal{R}]}$$

Equivalence of Shot Noise and Johnson Noise

The following derivation illustrates that shot noise and Johnson noise are not two independent noise mechanisms. In fact, we will show that both are the same for the special case of an ideal p-n diode under zero bias. At zero bias the photodetector can also be modeled as a resistor. Therefore the expression for Johnson noise should apply:

$$\langle i^2 \rangle = \frac{4kT\Delta f}{R}$$

The resistance of a photodiode with is

$$I = I_s(e^{V_a/V_t} - 1)_{is}$$

$$R = \frac{1}{\frac{dI}{dV_a}} = \frac{V_t}{I_s e^{V_a/V_t}}$$

or for zero bias, the Johnson noise current is given by:

$$\langle i^2 \rangle = \frac{4kT\Delta f I_s}{V_t} = 4I_s q\Delta f$$

whereas the shot noise current at $V_a = 0$ is given by:

$$\langle i^2 \rangle = 2qI_s(1 + e^{V_a/V_t}) = 4q\,I_s\,\Delta f$$

where we added the noise due to the diffusion current to the noise due to the (constant) drift current, since both noise mechanisms do not cancel each other. Equations are identical, thereby proving the equivalence between shot noise and Johnson noise in a photodiode at zero voltage. Note that this relation does not apply if the current is dominated by trap-assisted recombination/generation in the depletion region because of the non-equilibrium nature of the recombination/generation process.

Examples.

For a diode current of 1μA, a bandwidth Δf of 1 GHz and a responsivity, R, of 0.2A/W, the noise current $\sqrt{\langle i^2 \rangle}$ equals 18 nA, corresponding to a minimum detectable power of 89 nW or -40.5 dBm. Johnson noise in a 50 W resistor, over a bandwidth Δf of 1 GHz, yields a noise current of 0.58 μA and P_{min} = 2.9 μW or -25.4 dBm.

If the diode current is only due to the optical power, or $I = P_{min}R$, *then*

$$P_{min} = \frac{2q\Delta f}{R} = -58 \text{ dBm}$$

The sensitivity for a given bandwidth can also be expressed as a number of photons per bit:

$$\#\text{photons/bit} = \frac{P_{min}}{h\nu\Delta f}$$

For instance, for a minimal power of -30 dBm and a bandwidth of 1 GHz, this sensitivity corresponds to 4400 photons per bit.

Noise Equivalent Power and ac Noise Analysis

Assume the optical power with average value P_0 is amplitude modulated with modulation depth, *m*, as described by:

$$P_{in} = P_0\left(1 + me^{j\omega t}\right)$$

The ac current (RMS value) in the photodiode with responsivity, R, is then

$$i_{ph} = \frac{mP_0}{\sqrt{2}}R$$

which yields as an equivalent circuit of the photodiode a current source $\sqrt{\langle i^2 \rangle}$ in parallel with a resistance, R_{eq}, where

R_{eq} is the equivalent resistance across the diode and $\sqrt{\langle i^2 \rangle}$ is the noise source, which is given by:

$$\sqrt{< i^2 >} = 2q\, I\, \Delta f (P_0 R + I_{eq})$$

where the equivalent dark current also includes the Johnson noise of the resistor, R_{eq}:

$$I_{eq} = (I_{dark} + \frac{2V_t}{R_{eq}})$$

The signal to noise ratio is then given by:

$$\frac{S}{N} = \frac{i_{ph}^2 R_{eq}}{< i^2 > R_{eq}} = \frac{m^2 P_0^2 R^2}{2q\Delta f (P_0 R + I_{eq})}$$

from the above equation one can find the required optical power P_0 needed to obtain a given signal to noise ratio, *S/N*:

$$P_0 = (S/N)\frac{2q\Delta f}{m^2 \mathcal{R}}\{1 + \sqrt{1 + \frac{I_{eq} m^2}{q\Delta f (S/N)}}\}$$

The noise equivalent power is now defined as the ac (RMS) optical power needed to obtain a signal-to-noise ratio of one for a bandwidth of 1 Hz or:

$$\text{NEP} = \frac{mP_0}{\sqrt{2}}(\Delta f = 1, S/N = 1) = \frac{\sqrt{2}q}{m\mathcal{R}}\{1 + \sqrt{1 + \frac{I_{eq} m^2}{q}}\}$$

We now consider two limiting case in which the NEP is either limited by the optical power or by the dark current.

For $\frac{I_{eq} m^2}{q} << 1$ it is the average optical power rather than the dark current which limits the NEP.

$$\text{NEP} = \frac{2\sqrt{2}q}{mR}$$

The noise equivalent power can also be used to calculate the ac optical power if the bandwidth differs from 1Hz from:

$$\frac{mP_0}{\sqrt{2}} = \text{NEP}\,\Delta f$$

where the noise equivalent power has units of W/Hz. However, the optical power is mostly limited by the dark current for which the expressions are derived below.

For $\frac{I_{eq}m^2}{q} >> 1$ it is the dark current (including the Johnson noise of the resistor) which limits the NEP

$$\text{NEP} = \frac{\sqrt{2qI_{eq}}}{R}$$

Again one can use the noise equivalent power to calculate the minimum detectable power for a given bandwidth:

$$\frac{mP_0}{\sqrt{2}} = \text{NEP}\sqrt{\Delta f}$$

where the noise equivalent power has now units of $\text{W}/\sqrt{\text{Hz}}$.

Switching of a P-i-n Photodiode

A rigorous solution for the switching time of a P-i-n photodiode starts from the continuity equations for electrons and holes:

$$\frac{\partial n}{\partial t} = \frac{1}{q}\frac{\partial J_n}{\partial x} - \frac{np - n_i^2}{n + p + 2n_i}\frac{1}{\tau_0} + g_{op}(x,t)$$

$$\frac{\partial p}{\partial t} = -\frac{1}{q}\frac{\partial J_p}{\partial x} - \frac{np - n_i^2}{n + p + 2n_i}\frac{1}{\tau_0} + g_{op}(x,t)$$

with

$$J_n = q\mu_n n\varepsilon + qD_p \frac{\partial n}{\partial x}$$

$$J_p = q\mu_p p\varepsilon + qD_n \frac{\partial n}{\partial x}$$

and the electric field is obtained from Gauss's law. For a P-i-n diode with generation only at $t = 0$ and neglecting recombination and diffusion these equations reduce to:

$$\frac{\partial n}{\partial t} = \frac{\partial n}{\partial x}\mu_n\varepsilon \text{ and } \frac{\partial p}{\partial t} = -\frac{\partial p}{\partial x}\mu_p\varepsilon$$

Where the electric field, ε, is assumed to be a constant equal to:

$$\varepsilon = \frac{\phi_i - V_a}{d}$$

replacing $n(x,t)$ by $n^*(x - v_{nt})$ and $p(x,t)$ by $p^*(x - v_{pt})$ yields $v_n = -\mu_n\varepsilon$ and $v_p = \mu_p\varepsilon$.

The carrier distributions therefore equal those at $t = 0$ but displaced by a distance $\mu_n\varepsilon t$ for holes and $-\mu_p\varepsilon t$ for electrons. The total current due to the moving charge is a displacement current which is given by:

$$I_{ph}(t) = \frac{dQ}{dt} = \iiint \frac{\rho}{d}\frac{dx}{dt}dV = \frac{A}{d}\int_0^d \rho v dx$$

$$J_{ph}(t) = q\frac{A}{d}\varepsilon\int_0^d (\mu_n n + \mu_p p)dx$$

$$J_{ph}(t) = q\frac{A(\phi_i - V_a)}{d^2}[\mu_n n + \mu_p p]$$

for $t < |d/v_n|$ and $t < |d/v_p|$. For a uniform carrier generation this reduces to:

$$I_{ph}(t) = \frac{qA(\phi_i - V_a)}{d^2}[\mu_n n_0^*(d - |v_n t|) + \mu_p p_0^*(d - |v_p t|)]$$

$$I_{ph}(t) = \frac{qA(\phi - V_a)}{d}[\mu_n n_0^*(1 - \frac{|v_n t|}{d}) + \mu_p p_0^*(d - \frac{|v_p t|}{d})]$$

In the special case where $v_n = v_p$ or $\mu_n = \mu_p$, the full width half maximum (FWHM) of the impulse response is:

$$\text{FWHM} = \frac{d}{|v_n|^2} = \frac{d^2}{2\mu_n(\phi - V_a)} = \frac{t_r}{2} \text{ with } t_r = \frac{d^2}{\mu_n(\phi - V_a)}$$

Note: Rule of thumb to convert a pulse response to –3 dB frequency:Assuming the photodiode response to be linear, the FWHM can be related to the half-power frequency by calculating the Fourier transform. For a gaussian pulse response (which also yields a gaussian frequency response) this relation becomes

$$f_{-3\text{dB}} = \frac{440 \text{ GHz}}{\text{FWHM (in ps)}}$$

Since the bandwidth depends on the transit time, which in turn depends on the depletion layer width, there is a tradeoff between the bandwidth and the quantum efficiency.

Solution in the Presence of Drift, Diffusion and Recombination

If we simplify the SHR recombination rate to n/τ and p/τ and assume a constant electric field and initial condition $n(x,0) = n_0$, the electron concentration can be obtained by solving the continuity equation, yielding:

$$n(x,t) = e^{-\alpha x} \sum_k e^{-\xi_{kn} x} B_k \sin \frac{k\pi x}{d}$$

where

$$B_k = \frac{2n_0 k\pi}{(k\pi)^2 + (\alpha d)^2}\left[1 - (-1)^k e^{\alpha d}\right]$$

with

$$\xi_{kn} = D_n\left(\alpha^2 + \left(\frac{k\pi}{d}\right)^2 + \frac{1}{\zeta}\right) \text{ and } \alpha = \frac{\varepsilon}{2V_t}$$

For this analysis we solved the continuity equation with $n(0,t) = n(L,t) = 0$ implying infinite recombination at the edges of the depletion region. The initial carrier concentration n_0 can also be related to the total energy which is absorbed in the diode at time $t = 0$:

$$n_0 = \frac{E_{pulse}}{E_{ph}Ad}$$

and the photo current is

$$I_{ph}(t) = \frac{qA\mu_n(\phi_i - V_a)}{d}\sum_k C_k \exp(-\xi_k t)$$

with C_k given by

$$C_k = \frac{2n_0 (k\pi)^2 [1-(-1)^k e^{\alpha d}][1-(-1)^k e^{-\alpha d}]}{[(\alpha d)^2 + (k\pi)^2]^2}$$

The above equations can be used to calculate the impulse response of a photodiode. Each equation must be applied to electrons as well as holes since both are generated within the diode. Typically electrons and holes have a different mobility, which results in two regions with different slopes. This effect is clearly visible in GaAs diodes as illustrated with the figure below.

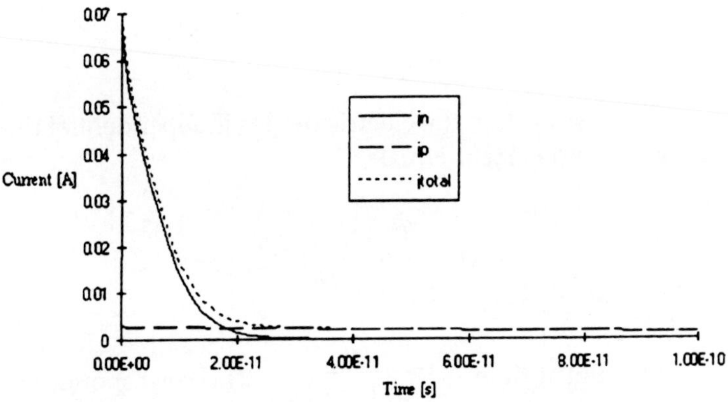

Fig. Photocurrent calculated using equation for a GaAs diode with $\phi_I - V_a = 0.3$ V, $E_{pulse} = 10^{-13}$ J, $E_{ph} = 2$ eV and d = 2μm.

Harmonic Solution

One can also solve the frequency response when illuminating with a photon flux $\Phi_1 e^{j\Omega t}$. If the photodiode has a linear response, both methods should be equivalent. To simplify the derivation, we assume that the total flux (in photons/s cm^2) is absorbed at $x = 0$. This is for instance the case for a p-i-n diode with a quantum well at the interface between the p-type and intrinsic region and which is illuminated with long wavelength photons, which only absorb in the quantum well. The carriers moving through the depletion region cause a conduction current, $J_{cond}(x)$:

$$J_{cond}(x) = q\Phi_1 e^{j\omega(t - x/v_n)}$$

where $v_n = \mu_n\,\varepsilon$ is a constant velocity.

From Ampere's law applied to a homogeneous medium, we find:

$$\frac{1}{\mu_0}\int \vec{B}d\vec{l} = \int \vec{J}d\vec{S} + \frac{\partial}{\partial t}\int \varepsilon_s \vec{E}d\vec{S}$$

And the total current is the sum of the conduction and the displacement current:

$$J_{total} = \frac{1}{d}\int_0^d [J_{cond} + \varepsilon_s \frac{\partial E}{\partial t}]dx$$

If we assume that the electric field is independent of time, the total photo current equals

$$J_{ph}(t) = \frac{q\phi_1\left(1 - e^{j\omega t_r}\right)e^{j\omega t_r}}{j\omega t_r}$$

with transit time $t_r = \dfrac{d^2}{\mu_n(\phi_i - V_a)}$. The corresponding –3 dB frequency is:

$$f - 3dB = \frac{2.78}{2\pi t_r} = \frac{443GH_z}{t_r[ps]}$$

Time Response Due to Carriers Generated in the Q.N. Region

For an infinitely long quasi-neutral (Q.N.) region and under stationary conditions, the generated carriers are only collected if they are generated within a diffusion length of the depletion region. The average time to diffuse over one diffusion length is the recombination time, τ. Postulating a simple exponential time response we find that the current equals

$$I_{ph(t)} = I_{ph}(0)e^{-t/t}$$

Because of the relatively long carrier lifetime in fast photodiodes, carriers absorbed in the quasi-neutral region produce a long "tail" in the pulse response and should be avoided.

Dynamic Range of a Photodiode

The dynamic range is the ratio of the maximal optical power which can be detected to the minimal optical power. In most applications the dynamic range implies that the response is linear as well. The saturation current, defined as the maximum current which can flow through the external circuit, equals:

$$I_{sat} = \frac{\phi_i - V_a}{R}$$

which yields an optimistic upper limit for the optical power:

$$P_{\max} = \frac{I_{sat}}{R} = \frac{\phi_i - V_a}{RR}$$

and the dynamic range is defined as the ratio of the maximum to the minimum power:

$$D.R. = \frac{P_{\max}}{P_{\min}}$$

Using for the minimum power the dynamic range becomes independent of the responsivity and equals:

$$D.R. = \frac{\phi_i - V_a}{R\sqrt{\langle i^2 \rangle}}$$

For example, if the equivalent noise current equals I_{eq} = 1μA, the bandwidth Δf = 1 GHz, the impedance R = 50 Ω, and the applied voltage V_a = 0, then the dynamic range equals 1.35 x 10^6 (for ϕ_i = 1.2 V) or 61.3 dB.

PHOTOCONDUCTORS

Photoconductors consist of a piece of semiconductor with two Ohmic contacts. Under illumination, the conductance of the semiconductor changes with the intensity of the incident optical power. The current is mainly due to majority carriers since they are free to flow across the Ohmic contacts. However the majority carrier current depends on the presence of the minority carriers. The minority carriers pile up at one of the contacts, where they cause additional injection of majority carriers until the minority carriers recombine. This effect can cause large "photoconductive" gain, which depends primarily of the ratio of the minority carrier lifetime to the majority carrier transit time. Long carrier lifetimes therefore cause large gain, but also a slow response time. The gain-bandwidth product of the photoconductor is almost independent of the minority carrier lifetime and depends only on the majority carrier transit time.

Consider now a photoconductor with length, *L*, width *W* and thickness *d*, which is illuminated a total power, *P*. The optical power, *P*(*x*), in the material decreases with distance due to absorption and is described by:

$$\frac{dP(x)}{dx} = -\alpha P(x), \text{ yielding } P(x) = P(0)\exp(-\alpha x)$$

The optical power causes a generation of electrons and holes in the material. Solving the diffusion equation for the steady state case and in the absence of a current density gradient one obtains for the excess carrier densities:

$$n' = p' = \frac{\zeta\alpha P(0)\exp(-\alpha x)}{h\zeta WL}$$

Where it was assumed that the majority carriers, which primarily contribute to the photocurrent, are injected from the contacts as long as the minority carriers are present. The photo current due to the majority carriers (here assumed to be n-type) is:

$$I_n = \frac{q}{h\zeta}(1-R)P_{in}\frac{\zeta}{t_r}(1-\exp(-\alpha d))$$

where t_r is the majority carrier transit time given by:

$$t_r = \frac{L^2}{\mu V}$$

The equation above also includes the power reduction due to the reflection at the surface of the semiconductor. The normalized photocurrent is plotted in Figure as a function of the normalized layer thickness for different ratio of lifetime to transit time.

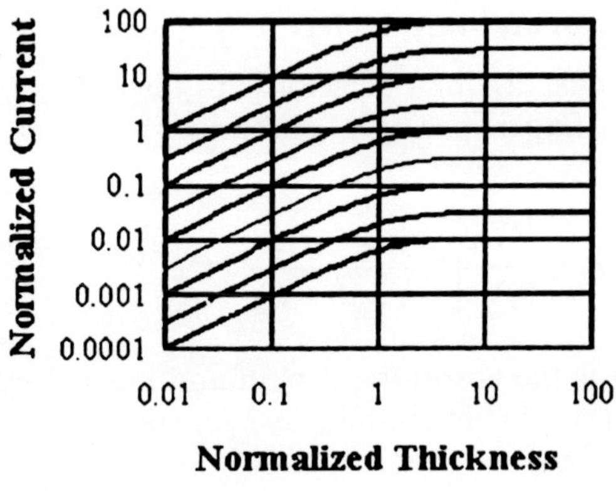

Fig. Normalized current $\frac{I_n}{P_{in}}\frac{h\tau}{q}$ versus normalized thickness ad as a function of the ratio of the minority carrier lifetime to the majority carrier transit time, t/t_r, ranging from 0.01 (bottom curve) to 100 (top curve)

As an example, consider a silicon photoconductor with $\mu_n = 1400$ cm^2/V-s and $\tau = 1$ µs. The photoconductor has a length of 10 micron and width of 100 micron. For an applied voltage of 5 Volt, the transit time is 143 ps yielding a photoconductive gain of 7000. For a normalized distance $\alpha d = 1$ and incident power of 1 mW the photocurrent equals 1.548 mA. A reflectivity of 30 per cent was assumed at the air/silicon interface.

High photoconductive gain is typically obtained for materials with a long minority carrier lifetime, τ, high mobility, μ_n, and above all a photoconductor with a short distance, L, between the electrodes.

METAL-SEMICONDUCTOR-METAL (MSM) PHOTODETECTORS

Metal-semiconductor-metal photodetectors are the simplest type of photodetectors since they can be fabricated with a single mask. They typically consist of a set of interdigitated fingers, resulting in a large active area and short distance between the electrodes.

Responsivity of an MSM Detector

The responsivity for a detector with thickness, d, surface reflectivity, R, finger spacing, L, and finger width, w, is given by:

$$\mathcal{R} = \left|\frac{I_{ph}}{P_{in}}\right| = \frac{q(1-R)L}{h\nu(L+w)}[1 - e^{-\alpha d}]$$

Where α is the absorption length and the reflectivity, R, of the air-semiconductor interface as a function of the incident angle is given by:

$$R_{TE} = \left(\frac{n_1 \cos\theta_i - n_2 \cos\theta_t}{n_1 \cos\theta_i + n_2 \cos\theta}\right)^2$$

$$R_{TM} = \left(\frac{n_2 \cos\theta_i - n_1 \cos\theta_t}{n_2 \cos\theta_i + n_1 \cos\theta}\right)^2$$

with $n_2 \sin\theta_t = n_t \sin\theta_i$

with θ_i the incident angle, and θ_t the transmitted angle. R_{TE} is the reflectivity if the electric field is parallel to the surface while R_{TM} is the reflectivity if the magnetic field is parallel to the surface. The reflectivity as a function of θ_i, for an air-GaAs interface is shown in Figure:

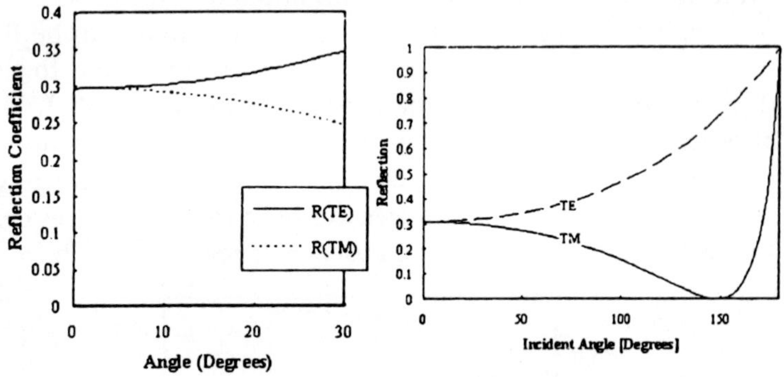

Fig. Angular dependencies of the reflectivity of an Air-to-GaAs interface

Including drift, diffusion and recombination the responsivity becomes:

$$\mathcal{R} = \left|\frac{I_{ph}}{P_{in}}\right| = \frac{q(1-R)L}{h\nu(L+w)}[1 - e^{-\alpha d}]\frac{V_a}{V_t}$$

$$\times\{[\frac{L_n^2}{L^3}(L - \frac{e^{(\beta_n - \alpha)L} - 1}{2(\beta_n - \alpha)}(1 - \coth(\beta_n L) + e^{\alpha L}\operatorname{csch}(\beta_n L)) + \frac{e^{-(\alpha+\beta_n)L} - 1}{2(\alpha+\beta_n)}(1 + \coth(\beta_n L) - e^{\alpha L}\operatorname{csch}(\beta_n L))]$$

$$[\frac{L_p^2}{L^3}(L - \frac{e^{(\beta_p - \alpha)L} - 1}{2(\beta_p - \alpha)}(1 - \coth(\beta_p L) + e^{\alpha L}\operatorname{csch}(\beta_p L)) + \frac{e^{-(\alpha+\beta_p)L} - 1}{2(\alpha+\beta_p)}(1 + \coth(\beta_p L) - e^{\alpha L}\operatorname{csch}(\beta_p L))]$$

with

$$\alpha = \frac{V_a}{2V_t L}, \quad \beta_n = \sqrt{\alpha^2 + \frac{1}{L_n^2}} \text{ and } \beta_p = \sqrt{\alpha^2 + \frac{1}{L_p^2}}$$

The above expression can be used to calculate the current as a function of the applied voltage. An example is shown in Figure. Both the electron and the hole current are plotted as is the total current. The difference between the electron and hole current is due to the recombination of carriers. For large voltages all photo-generated carriers are swept out yielding a saturation of the photocurrent with applied voltage, whereas for small voltages around zero diffusion is found to be the dominant mechanism. The ratio of the transit time to the diffusion time determines the current around zero volt. In the absence of velocity saturation botlr transit times depend on the carrier mobility so that the ratio becomes independent of the carrier mobility. This causes the *I-V* curves to be identical for electrons and holes in the absence of recombination.

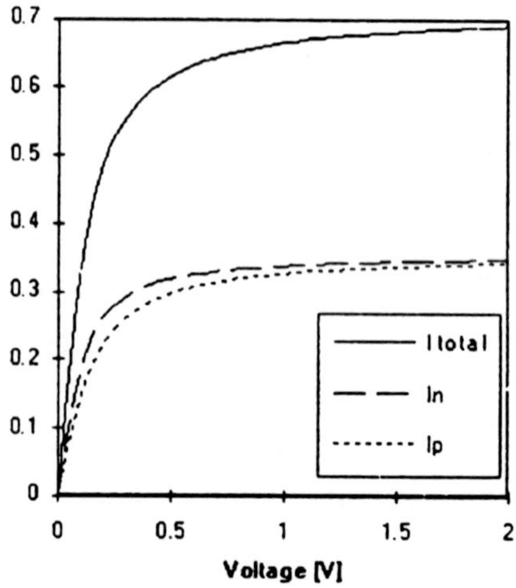

Fig. Current - Voltage characteristic of an MSM photodiode.

Pulse Response of an MSM Detector

The pulse response can be calculated by solving the time dependent continuity equation, yielding:

$$I_{ph}(t) = \frac{qAV_a}{L}\frac{(1-R)L}{L+w}[1-e^{-\alpha d}]\sum_k C_k[\mu_n e^{-\xi_{kn}t} + \mu_p e^{-\xi_{kp}t}]$$

with C_k given by:

$$C_k = \frac{2n_0 (k\pi)^2 [1-(-1)^k e^{\alpha d}][1-(-1)^k e^{-\alpha d}]}{[(\alpha d)^2 + (k\pi)^2]^2}$$

where

$$n_0 = \frac{E_{pulse}}{E_{ph} AL}$$

and

$$\xi_{kn} = D_n \left\{ \alpha^2 + \left(\frac{k\pi}{d} \right)^2 \right\} + \frac{1}{\zeta}, \xi_{kp} = D_p \left\{ \alpha^2 + \left(\frac{k\pi}{d} \right)^2 \right\} + \frac{1}{\xi}$$

and $\alpha = \dfrac{\varepsilon}{2V_t}$

This solution is plotted in Figure.

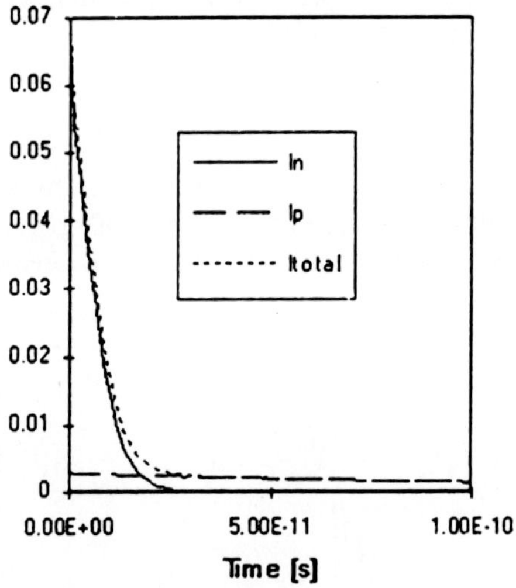

Fig. Transient behaviour (Pulse energy, E_{pulse} = 0.1 pJ, V_a = 0.3V)

Equivalent Circuit of an MSM Detector

The equivalent circuit of the diode consists of the diode capacitance, C_p, a parallel resistance, R_p, obtained from the slope of the *I*-V characteristics at the operating voltage in parallel to the photocurrent, I_{ph}, which is obtained by calculating the convolution of the impulse response and the optical input signal. A parasitic series inductance, L_B, primarily due to the bond wire, and a series resistance, R_s, are added to complete the equivalent circuit shown in Figure.

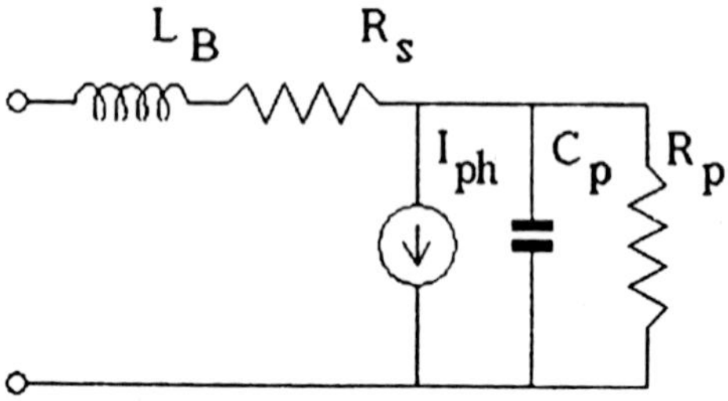

Fig. Equivalent circuit of an MSM detector

SOLAR CELLS

Solar cells are p-i-n photodiodes, which are operated under forward bias. The intention is to convert the incoming optical power into electrical power with maximum efficiency

THE SOLAR SPECTRUM

The solar spectrum is shown in Figure.The spectrum as seen from a satellite is referred to as the AM0 spectrum (where AM stands for air mass) and closely fits the spectrum of a black body at 5800 K. The total power density is 1353 W/m^2.

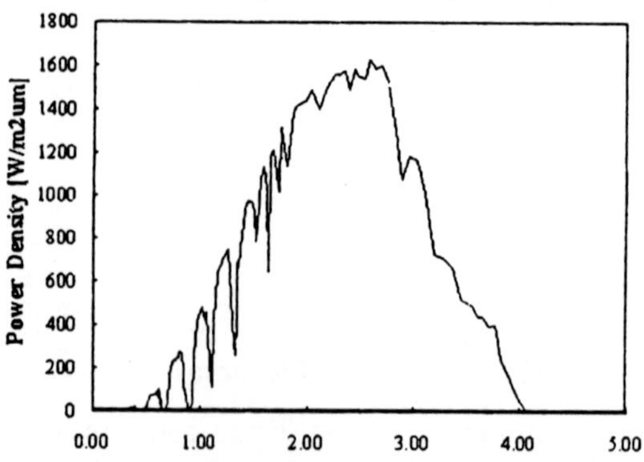

Fig. The solar spectrum under of AM1 conditions

The solar spectrum as observed on earth is modified due to absorption in the atmosphere. For AM1 (normal incidence) the power density is reduced to 925 W/cm^2 whereas for AM1.5 (45° above the horizon) the power density is 844 W/m^2. The irregularities in the spectrum are due to absorption at specific photon energies. The corresponding cumulative photocurrent is presented in Figure as a function of the photon energy.

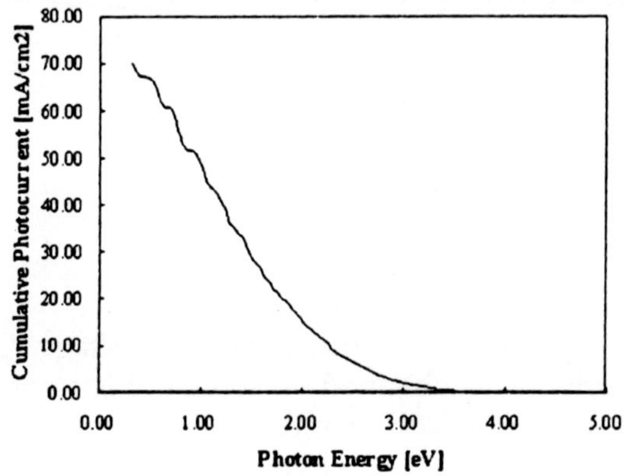

Fig. Cumulative Photocurrent versus Photon Energy under AM1 conditions

CALCULATION OF MAXIMUM POWER

The current through the solar cell can be obtained from:

$$I = I_s (e^{V_a / V_t} - 1) - I_{ph}$$

where I_s is the saturation current of the diode and I_{ph} is the photo current (which is assumed to be independent of the applied voltage V_a). This expression only includes the ideal diode current of the diode, thereby ignoring recombination in the depletion region. The short circuit current, I_{sc}, is the current at zero voltage which equals $I_{sc} = -I_{ph}$. The open circuit voltage equals:

$$V_{oc} = V_a (I = 0) = V_t \ln(\frac{I_{ph}}{I_s} + 1) \cong V_t \ln \frac{I_{ph}}{I_s}$$

The total power dissipation is then:

$$P = V_a I = I_s V_a (e^{V_a/V_t} - 1) - I_{ph} V_a$$

The maximum power occurs at $dP/dV_a = 0$. The voltage and current corresponding to the maximal power point are V_m and I_m.

$$\frac{dP}{dV_a} = 0 = I_s (e^{V_m/V_t} - 1) - I_{ph} + \frac{I_s V_m}{V_t} e^{V_m/V_t}$$

This equation can be rewritten as:

$$V_m = V_t \ln\left[\frac{I_{ph} + I_s}{I_s} \frac{1}{1 + \frac{V_m}{V_t}}\right] \cong V_t\left[\frac{V_{oc}}{V_t} - \ln\left(1 + \frac{V_m}{V_t}\right)\right]$$

by using equation for the open circuit voltage V_{oc}. A more accurate solution is obtained by solving this transcendental equation and substituting into equations. The maximum power can be approximated by:

$$P_m = I_m V_m \cong -I_{ph}\left(1 - \frac{V_t}{V_m}\right)\left(V_{oc} - V_t \ln\left(1 + \frac{V_m}{V_t}\right)\right)$$

$$P_m \cong -I_{ph}\left(V_{oc} - V_t \ln\left(1 + \frac{V_m}{V_t}\right) - \frac{V_{oc} V_t}{V_m}\right)$$

or

$$P_m = -I_{ph} \frac{E_m}{q}$$

where

$$E_m = q\left(V_{oc} - V_t In\left(1 + \frac{V_m}{V_t}\right) - \frac{V_{oc} V_t}{V_m}\right)$$

The energy E_m is the energy of one photon, which is converted to electrical energy at the maximum power point. The total photo current is calculated as (for a given bandgap E_g)

$$J_{ph}\left(E_g\right) = q$$

and the efficiency equals:

$$\eta = \frac{P_m}{P_{in}} = \frac{I_{ph} E_m}{q P_{in}}$$

CONVERSION EFFICIENCY FOR MONOCHROMATIC ILLUMINATION

This first order model provides an analytic approximation for the efficiency of a solar cell under monochromatic illumination.

$$\eta = \frac{P_m}{P_{in}} = \frac{I_{ph}}{P_{in}} \left(V_{oc} - V_t In \left(1 + \frac{V_m}{V_t} \right) - \frac{V_{oc} V_t}{V_m} \right)$$

and replace V_{oc} by the largest possible open circuit voltage, E_g/q, yielding:

$$V_m = \frac{E_g}{q} - V_t In \left(1 + \frac{E_g}{kT} \right)$$

and

$$\eta = \frac{I_{ph} V_m}{P_{in}} = \left[1 - \frac{2kT}{E_g} In \left(1 + \frac{E_g}{kT} \right) \right]$$

for a GaAs solar cell at 300K, E_g/kT = 55 so that the efficiency equals η = 85per cent

EFFECT OF DIFFUSION AND RECOMBINATION IN A SOLAR CELL

Photo Current Versus Voltage

$$0 = D_n \frac{d^2 n}{dx^2} + \mu_n \varepsilon \frac{dn}{dx} - \frac{n}{\zeta} + g_{op}$$

as well as a similar equation for holes. The photo current is obtained from

$$I_{ph}(V_a) = \frac{qA}{d^2} (\phi_i - V_a) \int_0^d (\mu_n n + \mu_p p) dx$$

Once this photocurrent is obtained the total current is obtained from:

$$I = I_s\left(e^{V_a/V_{t-1}}\right) - 1_{ph}(V_a)$$

To obtain the corresponding maximum power one has to repeat the derivation.

SPECTRAL RESPONSE

Because of the wavelength dependence of the absorption coefficient one expects the shorter wavelengths to be absorbed closer to the surface while the longer wavelengths are absorbed deep in the bulk. Surface recombination will therefore be more important for short wavelengths while recombination in the quasi-neutral region is more important for long wavelengths.

INFLUENCE OF THE SERIES RESISTANCE

$$V_{ext} = V_a + IR_s$$

$$I = I_s\left(e^{V_a/V_t} - 1\right) - 1_{ph}$$

$$P = V_{ext} I$$

Repeating the derivation, one can show that the maximum power condition is given by the following set of transcendental equations:

$$V_m = V_t \ln[\frac{I_{ph} + I_s}{I_s} \frac{1}{1 + \frac{V_m}{V_t} + \frac{2I_m R_s}{V_t}}]$$

$$I_m = I_s\left(e^{v_m/V_t} - 1\right) - I_{ph}$$

while the maximum external power equals: $P_{m,ext} = I_m (V_m + I_m R_s)$

LEDS

A light emitting diode consists of a p-n diode, which is designed so that radiative recombination dominates.

Homojunction p-n diodes, heterojunction p-i-n diodes where the intrinsic layer has a smaller bandgap (this structure is also referred to as a double-hetero-structure) and p-n diodes with a quantum well in the middle are all used for LEDs. We will only consider the p-n diode with a quantum well because the analytical analysis is more straightforward and also since this structure is used often in LEDs and even more frequently in laser diodes.

RATE EQUATIONS

The LED rate equations are derived from the continuity equations as applied to the p-n diode:

$$\frac{\partial n}{\partial t} = \frac{1}{q}\frac{\partial J_n}{\partial x} - R + G$$

where G is the generation rate per unit volume and R is the recombination rate per unit volume. This equation is now simplified by integrating in the direction perpendicular to the plane of the junction. We separate the integral in two parts: one for the quantum well, one for the rest of the structure.

$$\int_{qw} \frac{\partial n}{\partial t} + \int_{p-n} \frac{\partial n}{\partial t} = \frac{J}{q} - \frac{J_{SHR}}{q} - \frac{J_{bb}}{q} - \sum_k \left(N_k P_k - N_{ik}^2 \right) B_k$$

$$- \frac{NP - N_{i1}^2}{N + P + 2N_{i1}} \frac{1}{\tau_{\pi r}}$$

where k refers to the quantum number in the well. If we ignore the carriers everywhere except in the quantum well and assume that only the first quantum level is populated with electrons/holes and that the density of electrons equals the density of holes, we obtain:

$$\frac{\partial N}{\partial t} = \frac{J}{q} - B_1 N^2 - \frac{N}{2\tau_{nr}} + \frac{S}{\tau_{ab}}$$

where the last term is added to include reabsorption of photons. The rate equation for the photon density including loss of photons due to emission (as described with the photon lifetime τ_{ph}) and absorption (as described with the photon absorption time τ_{ab}) equals:

$$\frac{\partial S}{\partial t} = B_1 N^2 - \frac{S}{\tau_{ph}} + \frac{S}{\tau_{ab}}$$

The corresponding voltage across the diode equals:

$$V_a = \frac{E_{g,qw}}{q} + V_t In\left[\left(e^{N/N_{c,qw}} - 1\right)\left(e^{N/N^*_{v,qw}} - 1\right)\right]$$

Where the modified effective hole density of states in the quantum well, $N_{v,qw}^*$, accounts for the occupation of multiple hole levels. The optical output power is given by the number of photons, which leave the semiconductor per unit time, multiplied with the photon energy:

$$P_{out} = h\nu \frac{S}{\tau_{ph}} A(1-R)\frac{\Theta_c^2}{4}$$

where A is the active area of the device, R is the reflectivity at the surface and Θ_c is the critical angle for total internal reflection

$$R = \left(\frac{n_1 - n_2}{n_1 + n_2}\right)^2 \text{ and } \Theta_c = \sin^{-1}\left(\frac{n_1}{n_2}\right)$$

The reflectivity and critical angle for a GaAs Air interface are 30 per cent and 16° respectively.

DC SOLUTION TO THE RATE EQUATIONS

The time independent solution (indicated with the subscript 0) in the absence of reabsorption is obtained from:

$$0 = \frac{J_0}{q} - BN_0^2 - \frac{N_0}{2\tau_{ph}}$$

$$0 = BN_0^2 - \frac{S_0}{\tau_{ph}}$$

where B is the bimolecular recombination constant. Solving these equations yields:

$$N_0 = \frac{1}{4B\tau_{nr}}[\sqrt{1+\frac{16\tau_{nr}^2 BJ_0}{q}}-1]$$

For small currents this reduces to: ($J << q/16\tau_{nr}^2 B$)

$$N_0 = \frac{2\tau_{nr}J_0}{q}$$

which indicates that SHR recombination dominates, whereas for large currents one finds: ($J >> q/16\tau_{nr}^2 B$)

$$N_0 = \sqrt{\frac{J_0}{qB}}$$

The dc optical output power is:

$$P0 = h\nu BN_0^2 A(1-R)$$

This expression explains the poor efficiency of an LED. Even if no non-radiative recombination occurs in the active region of the LED, most photons are confined to the semiconductor because of the small critical angle. Typically, only a few percent of the photons generated escape the semiconductor. This problem is most severe for planar surface emitting LEDs. Better efficiencies have been obtained for edge emitting, "super luminescent" LEDs (where stimulated emission provides a larger fraction of photons which can escape the semiconductor) and LEDs with curved surfaces.

AC SOLUTION TO THE RATE EQUATIONS

We now assume that all variables can be written as a sum of a time independent term and a time dependent term (note that $n(t)$ is still a density per unit area):

$$N = N_0 + n_1(t) \quad J = J_0 + j_1(t)$$

$$S = S_0 + s_1(t) \quad P = P_0 + p_1(t)$$

$$V_a = V_{a,0} + v_a(t)$$

The rate equations for the time-dependent terms is then given by:

$$\frac{\partial n_1}{\partial t} = -B2N_0 n_1(t) - Bn_1^2(t) - \frac{n_1(t)}{2\tau_0} + \frac{j_1(t)}{q}$$

$$\frac{\partial s_1}{\partial t} = B2N_0 n_1(t) + Bn_1^2(t) - \frac{s_1(t)}{\tau_{ph}}$$

Next, we assuming the ac current of the form $j_1 = j_{1,0}e^{j\Omega t}$ and ignore the higher order terms. This results in a harmonic solution of the form:

$$n_1 = n_1, 0e^{j\omega t} \; s_1 = s_1, 0e^{j\omega t} \; p_1 = p_1, oe^{j\omega t}$$

with:

$$s_{1,0} = \frac{1}{q}\frac{B_2 N_0 j_1, 0\tau_{ph}\tau_{eff}}{q\left(1 + j\omega\tau_{ph}\right)\left(1 + j\omega\tau_{eff}\right)}$$

where τ_{eff} depends on N_0 as:

$$\tau_{eff} = \frac{1}{2BN_0 + \frac{1}{2\tau_0}}$$

and the ac responsivity is:

$$\frac{p_{1,0}}{j_{i,0}} = \frac{hv}{q}(1 - R)\frac{\pi\Theta_c^2}{4\pi}\frac{B2N_0\tau_{eff}}{\left(1 + j\omega\tau_{ph}\right)\left(1 + j\omega\tau_{eff}\right)}$$

At $\Omega = 0$ this also yields the differential quantum efficiency (D.Q.E)

$$D.Q.E. = \frac{p_{1,0}}{j_{i,0}}\frac{q}{hv} = \frac{(1 - R)\Theta_c^2 BN_0\tau_{eff}}{2} = \frac{(1 - R)\Theta_c^2 BN_0\tau_0}{4BN_0\tau_0 + 1}$$

EQUIVALENT CIRCUIT OF AN LED

The equivalent circuit of an LED consists of the p-n diode current source parallel to the diode capacitance and in series with a linear series resistance, *R*. The capacitance, *C*, is obtained from:

$$\frac{1}{C} = \frac{dV_a}{dQ} = \frac{dV_a}{qdN}$$

$$\frac{1}{C} = \frac{1}{q}V_t \frac{\left[\frac{1}{N_c} eN/N_c\left(e^{N/N_v-1}\right) + \frac{1}{N_v} e^{N/N_v\left(e^{N}/N_c-1\right)}\right]}{\left(e^{N/N_c}-1\right)\left(e^{N/N_v}-1\right)}$$

or

$$C = \frac{qN_0}{mV_t}$$

with

$$m = \frac{N_0 e^{N/N_c}}{N_c\left(e^{N/N_c}-1\right)} + \frac{N_0 e^{N/N_v}}{N_v\left(e^{N/N_{v-1}}\right)}$$

for $N_0 \ll N_c$ and/or N_v, $m = 2$ while for $N_0 \gg N_c$ and/or N_v,

$$m = \frac{N_0\left(N_c + N_v\right)}{N_c N_v}.$$

LASER DIODES

INTRODUCTION

Laser structure and principle of operation

Laser diodes consist of a p-n diode with an active region where electrons and holes recombine resulting in light emission. In addition, a laser diode contains an optical cavity where stimulated emission takes place. The laser cavity consists of a waveguide terminated on each end by a mirror. As an example, the structure of an edge-emitting laser diode is shown in Figure. Photons, which are emitted into the waveguide, can travel back and forth in this waveguide provided they are reflected at the mirrors. The distance between the two mirrors is the cavity length, labeled *L*.

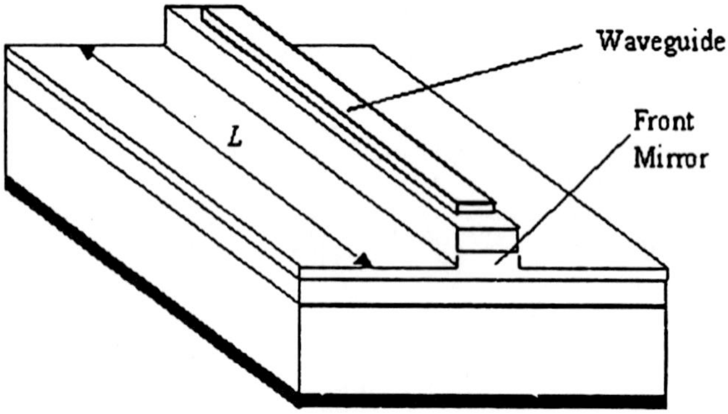

Fig. Structure of an edge-emitting laser diode.

Stimulated Emission and Modal Gain

The light in the waveguide is amplified by stimulated emission. Stimulated emission is a process where a photon triggers the radiative recombination of an electron and hole thereby creating an additional photon with the same energy and phase as the incident photon. This process is illustrated with Figure.This "cloning" of photons results in a coherent beam.

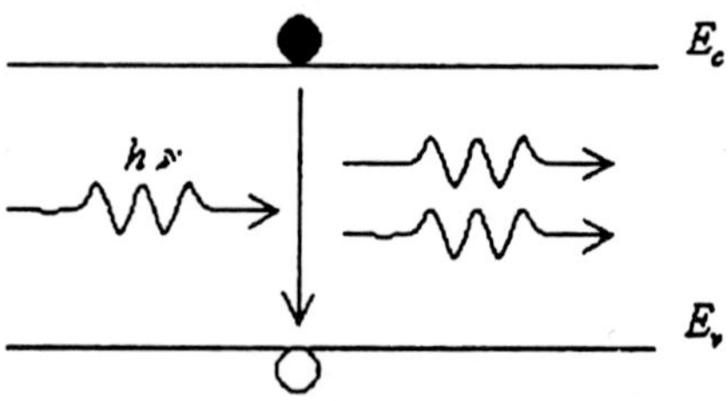

Fig. Stimulated emission of a photon.

The stimulated emission process yields an increase in photons as they travel along the waveguide.

Lasing condition

Combined with the waveguide losses, stimulated emission yields a net gain per unit length, *g*. The number of photons can therefore be maintained if the roundtrip amplification in a cavity of length, *L*, including the partial reflection at the mirrors with reflectivity R_1 and R_2 equals unity.

This yields the following lasing condition:

Roundtrip amplification = $e^{2gL}R_1R_2 = 1$

If the roundtrip amplification is less than one then the number of photons steadily decreases. If the roundtrip amplification is larger than one, the number of photons increases as the photons travel back and forth in the cavity and no steady state value would be obtained. The gain required for lasing therefore equals:

$$g = \frac{1}{2L} In \frac{1}{R_1 R_2}$$

Initially, the gain is negative if no current is applied to the laser diode as absorption dominates in the waveguide. As the laser current is increased, the absorption first decreases and the gain increases.

Output Power

The current for which the gain satisfies the lasing condition is the threshold current of the laser, I_{th}. Below the threshold current very little light is emitted by the laser structure. For an applied current larger than the threshold current, the output power, P_{out}, increases linearly with the applied current as illustrated with Figure. The output power therefore equals:

$$P_{out} = \eta \frac{hv}{q}(I - I_{th})$$

where hv is the energy per photon. The factor, η, indicates that only a fraction of the generated photons contribute to the output power of the laser as photons are partially lost through the other mirror and throughout the waveguide.

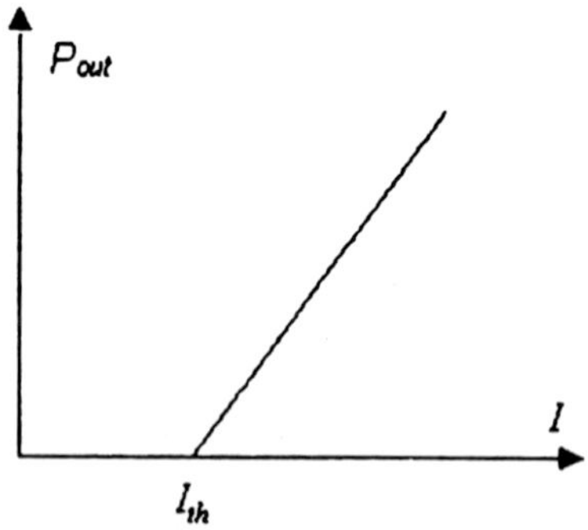

Fig. Output power from a laser diode versus the applied current.

LASER CAVITIES AND LASER CAVITY MODES

A laser diode consists of a cavity, defined as the region between two mirrors with reflectivity R_1 and R_2, and a gain medium, in our case a quantum well. The optical mode originates in spontaneous emission, which is confined to the cavity by the waveguide. This optical mode is amplified by the gain medium and partially reflected by the mirrors. The modal gain depends on the gain of the medium, multiplied with the overlap between the gain medium and the optical mode which we call the confinement factor, Γ, or:

$$\text{modal gain} = g(N)\Gamma$$

This confinement factor lasing occurs when the round trip optical gain equals the losses. For a laser with modal gain $g(N)\Gamma$ and waveguide loss, α, this condition implies:

$$R_1R_2\exp[2(g(N) - \alpha)L] = 1$$

where L is the length of the cavity. The distributed loss of the mirrors is therefore:

$$\text{mirrorloss} = \frac{1}{L} In \frac{1}{\sqrt{R_1R_2}}$$

Longitudinal Modes in the Laser Cavity

Longitudinal modes in the laser cavity correspond to standing waves between the mirrors. If we assume total reflection at the mirrors this wave contains $N/2$ periods where

N is an integer. For a given wave length λ and a corresponding effective index, n_{eff}, this yields:

$$N = \frac{2n_{eff}L}{\lambda}$$

Because of dispersion in the waveguide, a second order model should also include the wavelength dependence of the effective index. Ignoring such dispersion effects, we find the difference in wavelength between two adjacent longitudinal modes from:

$$N = \frac{2n_{eff}L}{\lambda_1}$$

$$N+1 = \frac{2n_{eff}L}{\lambda_2}$$

$$\Delta\lambda = 2Ln_{eff}\left(\frac{1}{N} - \frac{1}{N+1}\right) \cong \frac{\lambda_1^2}{2Ln_{eff}}$$

Longer cavities therefore have closer spaced longitudinal modes. An edge emitting (long) cavity with length of 300 μm, n_{eff} = 3.3, and λ = 0.8 μm has a wavelength spacing Dl of 0.32 nm while a surface emitting (short) cavity of 3 μm has a wavelength spacing of only 32 nm. These wavelength differences can be converted to energy differences using:

$$\Delta E = \frac{hc}{2Ln_{eff}} = E_{ph}\frac{\Delta l}{\lambda_2}$$

so that 0.32 nm corresponds to –6.2 meV and 32 nm to 620 meV. A typical width of the optical gain spectrum is 60 meV, so that an edge emitter biased below threshold can easily contain 10 longitudinal modes, while for a surface emitter the cavity must be carefully designed so that the longitudinal mode overlaps with the gain spectrum.

A more detailed analysis of a Fabry-Perot etalon is described in Appendix 15, providing the reflectivity, absorption and transmission as a function of photon energy.

Waveguide Modes

The optical modes in the waveguide determine the effective index needed to calculate the longitudinal modes as well as the confinement factor. Starting from Maxwell's equations in the absence of sources:

$$\bar{\nabla} \times \bar{H} = \varepsilon_0 n^2 (x, y, z)$$

$$\bar{\nabla} \times \bar{\varepsilon} = \mu_0 \frac{d\bar{H}}{\partial t}$$

and assuming a propagating wave in the *z*-direction and no variation in the *y*-direction, we obtain the following one-dimensional reduced wave equation for a time harmonic field, $\varepsilon = \varepsilon_x e^{j\omega t}$, of a TM mode:

$$\frac{d^2 \varepsilon_x}{dx^2} + \left(n^2 (x) k^2 - \beta^2\right) \varepsilon_x = 0$$

Where β, is the propagation constant given by $\beta = \frac{\omega}{c} n_{eff}$, and $k = \frac{\omega}{c}$. The equation then becomes:

$$\frac{d^2 \varepsilon_x}{dx^2} + \left(n^2 (x) - n_{eff}^2\right) \varepsilon_x = 0$$

this equation is very similar to the Schrödinger equation. In fact, previous solutions for quantum wells can be used to solve Maxwell's equation by setting the potential $V(x)$ equal to $-n^2 (x)$ and replacing $\frac{\omega^2}{c^2}$ by $\frac{2m^*}{\hbar^2}$. The energy eigenvalues, E_l, can then be interpreted as minus the effective indices of the modes: $-n^2_{eff,l}$. One particular waveguide of interest is a slab waveguide consisting of a piece of high refractive index material, n_1, with thickness *d*, between two infinitely wide cladding layers consisting of lower refractive index material, n_2.

$$V_0 = -n_2^2 + n_1^2 E_{10} = \frac{c^2}{\omega^2}\frac{\pi^2}{d^2}$$

or,

$$d \le \frac{c}{\omega}\pi\frac{1}{\sqrt{n_1^2 - n_2^2}} = \frac{\lambda}{2\sqrt{n_1^2 - n_2^2}}$$

For $\lambda = 0.8$ μm, $n_1 = 3.5$ and $n_2 = 3.3$ one finds $d \le 0.34$ μm.

The Confinement Factor

The confinement factor is defined as the ratio of the modal gain to the gain in the active medium at the wavelength of interest:

$$\Gamma = \frac{\text{modal gain}}{g} = \frac{\int_{-\infty}^{\infty} g(x)\,|\varepsilon_x|^2\,dx}{\int_{-\infty}^{\infty} |\varepsilon_x|^2\,dx}$$

for a quantum well with width L_x, the confinement factor reduces to

$$\Gamma = \frac{\int_{-L_x/2}^{L_x/2} |\varepsilon_x|^2\,dx}{\int_{-\infty}^{\infty} |\varepsilon_x|^2\,dx}$$

which is approximately ≅ 0.02...0.04 for a typical GaAs single quantum well laser

EMISSION, ABSORPTION AND MODAL GAIN

The analysis of a semiconductor laser diode requires a detailed knowledge of the modal gain, which quantifies the amplification of light confined to the lasing mode. To find the modal gain, one starts from the requirement that the emission as well as absorption of photons, must conserve both energy and momentum of all particles involved in the process. The conservation of energy requires that the photon energy equals the difference between the electron and hole energy:

$$E_{ph} = E_n - E_p$$

with

$$E_n = E_c + E_{1n} + \frac{\hbar^2 k_n^2}{2m_n^*}$$

$$E_p = E_v - E_{1p} + \frac{\hbar^2 k_p^2}{2m_p^*}$$

The conservation of momentum requires that the electron momentum equals that of the empty state it occupies in the valence band plus the momentum of the photon

$$k_n = k_p + k_{ph}$$

The photon momentum is much smaller than that of the electron and hole, so that the electron and hole momentum are approximately equal. As a result we can replace k_n and k_p by a single variable k. Equations then result in:

$$E_{ph} = E_{g,qwl} + \frac{\hbar^2 k^2}{2m_r^*}$$

where $E_{g,qw1}$ is the energy between the lowest electron energy in the conduction band and the lowest hole energy in the valence band. m_r^* is the reduced effective mass given by:

$$\frac{1}{m_r^*} = \frac{1}{m_n^*} + \frac{1}{m_p^*}$$

The electron and hole energies, E_n and E_p, can then be expressed as a function of the photon energy by:

$$E_n = E_c + E_{1n} + \left(E_{ph} - E_{g,qwl}\right)\frac{m_r^*}{m_n^*}$$

$$E_p = E_v - E_{lp} - \left(E_{ph} - E_{g,qwl}\right)\frac{m_r^*}{m_p^*}$$

The emission and absorption spectra ($\beta(E_{ph})$ and $\alpha(E_{ph})$) of a quantum well depend on the density of states and the occupancy of the relevant states in the conduction and valence band. Since the density of states in the conduction and valence band is constant in a quantum well, the emission and absorption can be expressed as a product of a maximum emission and absorption rate and the probability of occupancy of the conduction and valence band states, namely:

$$\beta\left(E_{ph}\right) = \beta_{\max} f_n\left(E_n\right)\left[1 - f_p\left(E_p\right)\right]$$

$$\alpha\left(E_{ph}\right) = \alpha_{\max}\left[1 - f_n\left(E_n\right)\right] f_p\left(E_p\right)$$

Stimulated emission occurs if an incoming photon triggers the emission of another photon. The net gain in the semiconductor is the stimulated emission minus the absorption. The maximum stimulated emission equals the maximum absorption since the initial and final states are simply reversed so that the transition rates as calculated based on the matrix elements are identical. The net gain is then given by:

$$g\left(E_{ph}\right) = \beta\left(E_{ph}\right) - \alpha\left(E_{ph}\right) = g_{\max}\left[f_n\left(E_n\right) - f_p\left(E_p\right)\right]$$

where the maximum stimulated emission and the maximum absorption were replaced by the maximum gain, g_{max}. The normalized gain spectrum is shown in Figure for different values of the carrier density. The two staircase curves indicate the maximum possible gain and the maximum possible absorption in the quantum well.

The theoretical gain curve of Figure exhibits a sharp discontinuity at $E_{ph} = E_{g,qw1}$. The gain can also be expressed as a function of the carrier densities, N and P, when assuming that only one electron and one hole level is occupied:

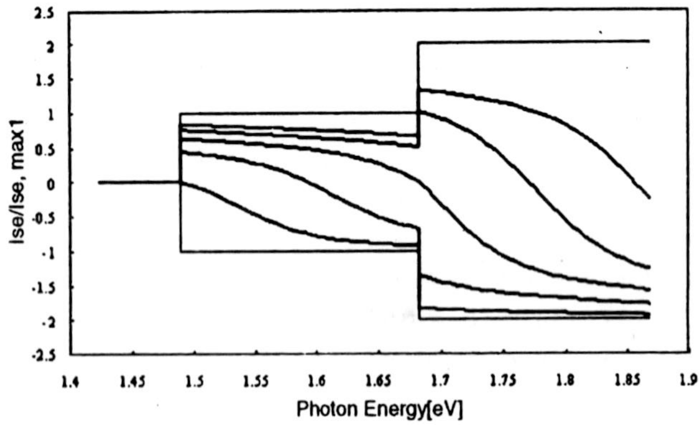

Fig. Normalized gain versus photon energy of a 10nm GaAs quantum well for a carrier density of 10^{12} (lower curve), 3 x 10^{12}, 5 x 10^{12}, 7 x 10^{12} and 9 x 1012 (upper curve) cm-2.

$$g(E_{ph}) = g_{\max}\left[\frac{1-\exp\left(-\frac{N}{N_{c,qw}}\right)}{1+\exp\left(-\frac{N}{N_{c,qw}}\right)\left[\exp\left(\frac{E_{ph}-E_{g,qwl}}{KT}\frac{m_r^*}{m_n^*}\right)-1\right]} - \frac{\exp\left(-\frac{P}{N_{v,qw}}\right)}{\left(1-\exp\left(-\frac{P}{N_{v,qw}}\right)\right)\exp\left(\frac{E_{g,qwl}-E_{ph}}{kT}\frac{m_r^*}{m_p^*}\right)+\exp\left(-\frac{P}{N_{v,qw}}\right)}\right]$$

The peak value at $E_{ph} = E_{g,qw1}$, assuming quasi-neutrality ($N = P$) is then:

$$g_{peak} = g\left(E_{g,qwl}\right) = g_{\max}\left(1 - e^{-N/N_c} - e^{-N/N_v}\right)$$

The maximum gain can be obtained from the absorption of light in bulk material: One can verify that the wavefunction of a free electron in bulk material is the same as the wavefunction in an infinite stack of infinitely deep quantum wells, provided the barriers are infinitely thin and placed at the nodes of the wavefunction. This means that for such a set of quantum wells the absorption would be the same as in bulk provided that the density of states is also the same. This is the case for $E_{ph} = E_{qw1}$, so that the maximum gain per unit length is given by:

$$g_{\max} = K\sqrt{E_{qwl} - E_1} = \frac{K}{2}\sqrt{\frac{h^2}{2m_r^*}\frac{1}{L_x}}$$

where L_x is the width of the quantum well. This expression shows that the total gain of a single quantum well due to a single quantized level is independent of the width. The corresponding value for GaAs quantum wells is 0.006 or 0.6per cent.

Experimental gain curves do not show the discontinuity at $E_{ph} = E_{qw1}$ due to inter-carrier scattering which limits the lifetime of carriers in a specific state. The line width of a single set of electron and hole levels widens as a function of the scattering time, which disturbs the phase of the atomic oscillator. Therefore, an approximation to the actual gain curve can be obtained by convoluting with a Lorenzian line shape function:

$$g(E_{ph}) = \int g_{\max}[F_n(E_n) - F_p(E_p)] \frac{\Delta\nu}{2\pi[(\nu - \frac{E_{ph}}{h})^2 + (\frac{\Delta\nu}{2})^2]} d\nu$$

with $\Delta\nu = \frac{1}{\pi}\frac{1}{\tau}$, where τ is the carrier collision time in the quantum well. The original and convoluted gain curves are shown in Figure.

For lasers with long cavities such as edge-emitter lasers, one finds that the longitudinal modes are closely spaced so that lasing will occur at or close to the peak of the gain spectrum. It is therefore of interest to find an expression for the peak gain as a function of the carrier density. A numeric solution is shown in Figure where the peak gain is normalized to the maximum value of the first quantized energy level. Initially, the gain peak is linear with carrier concentration but saturates because of the constant density of states, until the gain peak associated with the second quantized level takes over.

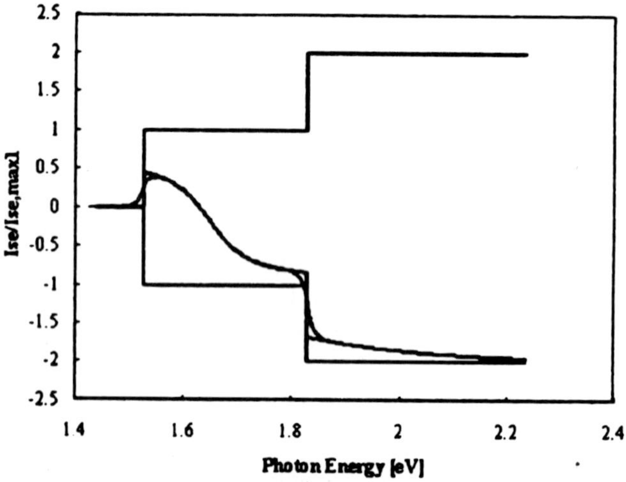

Fig. Original and convoluted gain spectrum of a 10 nm GaAs quantum well with a carrier density of 3×10^{12} cm^{-2} and a collision time of 0.09 ps.

Since the peak gain will be relevant for lasing, we will consider it more closely. As a first order approximation we will set the peak gain $g(N)$ equal to:

$$g(N) = l(N - N_{tr})$$

where l is the differential gain coefficient. This approximation is only valid close to $N = N_{tr}$, and even more so for quantum well lasers as opposed to double-heterostructure lasers. An approximate value for the differential gain coefficient of a quantum yielding:

$$l = g_{\max}\left[\frac{e^{-N_{tr}/N_c}}{N_c} + \frac{e^{-N_{tr}/N_c}}{N_v}\right]$$

From Figure one finds that the material becomes "transparent" when the gain equals zero or:

$$g(N_{tr}) = 0 = g_{\max}\left[F_n(E_n)\left(1 - F_p(E_p)\right) - \left(1 - F_n(E_n)\right)F_p(E_p)\right]$$

which can be solved yielding:

$$E_{ph} = E_n - E_p = E_{Fn} - E_{Fp} = qV_a$$

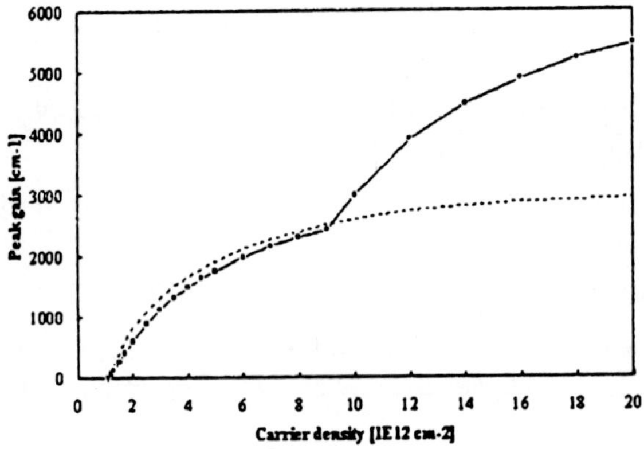

Fig. Calculated gain versus carrier density for a 10 nm GaAs quantum well (solid line) compared to equation

The transparency current density is defined as the minimal current density for which the material becomes transparent for any photon energy larger than or equal to $E_{g,qw1}$. This means that the transparency condition is fulfilled for $V_a = \frac{E_{g.qwl}}{q}$.

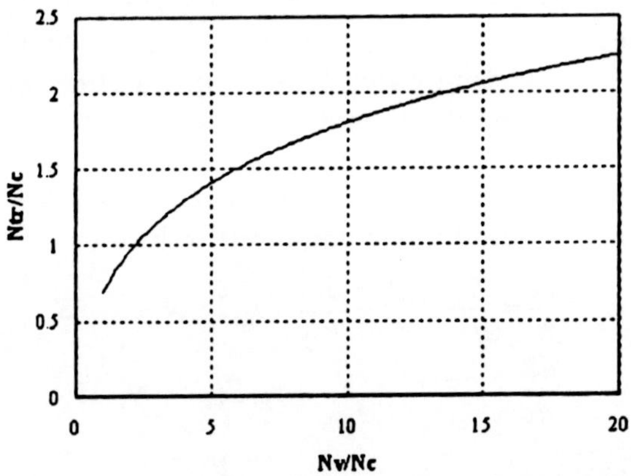

Fig. Normalized transparency carrier density versus the ratio of the effective density of states in the valence and conduction band.

The corresponding carrier density is referred to as N_{tr}, the transparency carrier density. The transparency carrier density can be obtained from by setting $g_{max} = 0$, yielding

$$N_{tr} = -N_c In\left(1 - e^{N_{tr}/N_c}\right)$$

This expression can be solved by iteration for $N_v > N_c$. The solution is shown in Figure.

To include multiple hole levels one simply replaces N_v by $N_{v,qw}^*$.

THE RATE EQUATIONS FOR A LASER DIODE.

Rate equations for each longitudinal mode, λ, with photon density S_λ and carrier density N_λ which couple into this mode are:

$$\frac{\partial N_\lambda}{\partial t} = \frac{J_\lambda}{q} - B_\lambda N_\lambda^2 - \frac{N_\lambda}{2\tau_0} + \sum_k \frac{N_\lambda}{\tau_{k,\lambda}} - \sum_k \frac{N_\lambda}{\tau_{k\lambda,k}} - \frac{\partial S_\lambda}{\partial x}\frac{\partial x}{\partial t}$$

$$\frac{\partial S_\lambda}{\partial t} = \beta_\lambda B_\lambda N_\lambda^2 - \frac{S_\lambda}{\tau_{ph,\lambda}} + \frac{\partial S_\lambda}{\partial x}\frac{\partial x}{\partial t}, \lambda = 1, 2, ..., \lambda_{MAX}$$

Rather than using this set of differential equations for all waveguide modes, we will only consider one mode with photon density S, whose photon energy is closest to the gain peak. The intensity of this mode will grow faster than all others and eventually dominate. This simplification avoids the problem of finding the parameters and coefficients for every single mode. On the other hand it does not enable to calculate the emission spectrum of the laser diode. For a single longitudinal mode the rate equations reduce to:

$$\frac{dN}{dt} = \frac{J}{q} - BN^2 - \frac{N}{2\tau_0} - v_{gr}\Gamma\ell(N - N_{tr})S$$

$$\frac{dS}{dt} = \beta BN^2 - \frac{S}{\tau_{ph}} + v_{gr}\Gamma\ell(N - N_{tr})S$$

$$P_1 = v_{gr}SW \ln\frac{1}{\sqrt{R_1}}$$

DC Solution to the Rate Equations

The time independent rate equations, ignoring spontaneous emission, are:

$$0 = \frac{J_0}{q} - BN_0^2 - \frac{NO}{2\tau_0} - Vgr\Gamma l\left(N_0 - N_{tr}\right)S_0$$

$$0 = \frac{S_0}{\tau_{ph}} + v_{gr}\Gamma l\left(N_0 - N_{tr}\right)S_0$$

where the photon life time is given by:

$$\frac{1}{\tau_{ph}} = \frac{1}{S}\frac{\partial x}{\partial t}\frac{\partial S}{\partial x} = vgr\left(\alpha + \frac{1}{L}In\frac{1}{\sqrt{R_1R_2}}\right)$$

from which we can solve the carrier concentration while lasing:

$$N_0 = N_{tr} + \frac{1}{\tau_{ph}v_{gr}\Gamma l}$$

which is independent of the photon density. The threshold current density is obtained when $S_0 = 0$

$$J_0 |_{(S_0=0)} = J_{th} = q\left(BN_0^2 + \frac{N_0}{2\tau_0}\right)$$

The photon density above lasing threshold, and power emitted through mirror R_1, are given by:

$$S_0 = \frac{J_0 - J_{th}}{q}\frac{1}{v_{gr}\Gamma\left(N_0 - N_{tr}\right)}$$

and the power emitted through mirror 1 is:

$$P_1, 0 = hvS_0Wv_{gr}In\frac{1}{\sqrt{R_1}}$$

The differential efficiency of the laser diode is:

$$D.E. = \frac{dP_1, 0}{dI_0} = \frac{hv}{q}\frac{In\frac{1}{\sqrt{R_1}}}{In\frac{1}{\sqrt{R_1 + R_2}} + \alpha L}$$

and the quantum efficiency is:

$$\eta = \frac{q}{h\nu}\frac{dP_{1,0}}{dI_0} = \frac{In\frac{1}{\sqrt{R_1}}}{In\frac{1}{\sqrt{R_1 + R_2}} + \alpha L}$$

Efficient lasers are therefore obtained by reducing the waveguide losses, increasing the reflectivity of the back mirror, decreasing the reflectivity of the front mirror and decreasing the length of the cavity. Decreasing the reflectivity of the mirror also increases the threshold current and is therefore less desirable. Decreasing the cavity length at first decreases the threshold current but then rapidly increases the threshold current.

AC Solution to the Rate Equations

Assuming a time-harmonic solution and ignoring higher order terms (as we did for the LED) the rate equations become:

$$j\omega n_1 = \frac{j_1}{q} - \frac{n_1}{\tau_{eff}} - v_{gr}\Gamma l(N_0 - N_{tr})s_1 - v_{gr}\Gamma \ln_l S_0$$

$$j\omega s_1 = -\frac{s_1}{\tau_{ph}} + v_{gr}\Gamma l(N_0 - N_{tr})s_1 + v_{gr}\Gamma \ln_l S_0$$

where τ_{eff} is the same as for an LED and given by equation. Using $v_{gr}\Gamma l(N_0 - N_{tr}) = \frac{1}{\tau_{ph}}$ these equations can be solved yielding:

$$j_1 = j\omega q n_1 + q n_1\left(\frac{1}{\tau_{eff}} + \Gamma l S_0 v_{gr}\right) + q n_1 \frac{\Gamma l S_0 v_{gr}}{j\omega\tau_{ph}}$$

We now replace n_1 by relating it to the small signal voltage v_1:

$$v_1 = \frac{m V_t n_1}{N_0}$$

The equation for the small signal current, I_1, can be then written as

$$i_1 = \left(j\omega C + \frac{1}{R} + \frac{1}{j\omega L} \right) v_1$$

with $C = \dfrac{qN_0 A}{mV_t}$, and $m = \dfrac{N_0 e^{N/N_c}}{N_c\left(e^{N/N_c} - 1\right)} + \dfrac{N_0 e^{N/N_v}}{N_v\left(e^{N/N_v} - 1\right)}$,

where A is the area of the laser diode.

$$\frac{1}{R} = C\left(\frac{1}{\tau_{eff}} + \Gamma l S_0 v_{gr} \right)$$

and

$$L = \frac{1}{C} \frac{\tau_{ph}}{\Gamma l S_0 v_{gr}}$$

Small Signal Equivalent Circuit

Adding parasitic elements and the circuit described by the equation we obtain the equivalent circuit of Figure, where L_B is a series inductance, primarily due to the bond wire, R_s is the series resistance in the device and C_p is the parallel capacitance due to the laser contact and bonding pad.

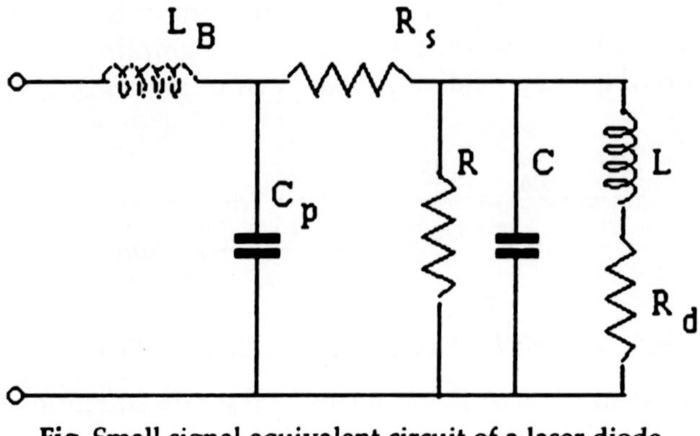

Fig. Small signal equivalent circuit of a laser diode

The resistor, R_d, in series with the inductor, L, is due to gain saturation and can be obtained by adding a gain saturation term to equation. The optical output power is proportional to the current through inductor L, $i_{1,L}$, which is given by:

$$i_{l,L} = \frac{qAs_1}{\tau_{ph}} = qAs_1V_{gr}\left(\alpha + \frac{1}{L} in \frac{1}{\sqrt{R_1 + R_2}}\right)$$

and the corresponding power emitted from mirror R_1

$$p_1 = s_1 hw_{gr} W In \frac{1}{\sqrt{R_1}}$$

When further ignoring the parasitic elements and the gain saturation resistance, R_d, one finds the ac responsivity p_1/I_1 as:

$$\frac{p_1}{i_1} = \frac{h\nu}{q} \frac{\ln\frac{1}{\sqrt{R_1}}}{\ln\frac{1}{\sqrt{R_1+R_2}} + \alpha L} \frac{1}{1 + j\omega\frac{L}{R} + (j\omega)^2 LC}$$

from which we find the relaxation frequency of the laser:

$$\omega_0 = \frac{1}{\sqrt{LC}} = \sqrt{\frac{\Gamma a S_0 v_{gr}}{\tau_{ph}}} = \sqrt{\frac{\Gamma a P_0}{\tau_{ph} h\nu W \ln\frac{1}{\sqrt{R_1}}}}$$

The relaxation frequency is therefore proportional to the square root of the DC output power. The amplitude at the relaxation frequency relative to that at zero frequency equals:

$$\frac{p_1|_{\omega=\omega_0}}{p_1|_{\omega=0}} = \frac{R}{L\omega_0} = \frac{1}{\frac{1}{\omega_0 \tau_{eff}} + \tau_{ph}\omega_0}$$

LARGE SIGNAL SWITCHING OF A LASER DIODE

Because of the non-linear terms in the rate equations the large signal switching of a laser diode exhibits some peculiar characteristics. The response to a current step is shown in the Figure. The carrier density initially increases linearly with time while the photon density remains very small since stimulated emission only kicks in for $N > N_0$.

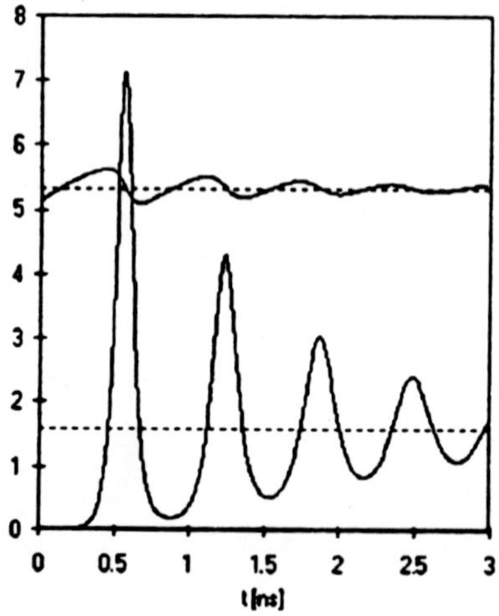

Fig. Optical power and normalized carrier concentration versus time when applying a step current at $t = 0$ from $I = 0.95\ I_{th}$ to $I = 1.3\ I_{th}$.

Both the carrier density and the photon density oscillate around their final value. The oscillation peaks are spaced by roughly $2\Pi/\omega_0$, where ω_0 is the small signal relaxation frequency at the final current. The photon and carrier densities are out of phase as carriers are converted into photons due to stimulated emission, while photons are converted back into electron-hole pairs due to absorption. One can use the non-linear behaviour to generate short optical pulses. By applying a current pulse, which is long enough to initiate the first peak in the oscillation, but short enough to avoid the second peak, one obtains a single optical pulse, which is significantly shorter that the applied current pulse. This method is referred to as gain switching or current spiking.